COSMOS

The Yearbook of the Traditional Cosmology Society

Volume 6
1990

CONTESTS

edited by
ANDREW DUFF-COOPER
General Editor of Cosmos: Emily Lyle

EDINBURGH UNIVERSITY PRESS

Advisory Board

© Edinburgh University Press 1990
22 George Square, Edinburgh

Set in Linotron Times Roman
by Photoprint Ltd, Torquay and
printed in Great Britain by
Page Bros Ltd, Norwich

ISBN 0 7486 0199 6 (paper)

Contents

ANDREW DUFF-COOPER

Introduction

The present *Yearbook* of the Traditional Cosmology Society derives from the Society's conference on Contests held at the University of Dundee, Scotland, in August 1989. Some essays that appear in this volume (those of Sharma, Lowry, Hendry, Kehoe, Davies, Ó Crualaoich, and Duff-Cooper) are more or less revised versions of papers given to the conference. The other papers published here were solicited from people who participated in the Dundee conference but read papers they did not wish to publish in the present collection and from people who did not take part in the conference. These writers were asked for a contribution in the main because it was judged that they would say something interesting and because they would expand the geographical and intellectual extent of the collection.

Wide geographical extension — the papers move from India (though for the sake of the conceit that Mr Sharma delivered his paper in Dundee we might say from there to India), east through Indonesia and China to Japan, across the Pacific to Mesoamerica, north to the northern United States and southern Canada, and then back to southern Europe and Ireland, returning finally to Scotland — does not stand in need of justification. It is well established what a worldwide purview brings to any humane study: among other benefits, an awareness of both the variety and the similarity, and of the relativity, of its subject-matter, and a liberation (referred to again below) from the narrow confines of small, sharply delineated parts of the world and equally constricting forms of life.

The intellectual extension of the chapters that make up this volume, though, might be thought by some to require rather more justification. The Traditional Cosmology Society is a forum for the international study of myth, religion, cosmology, and related matters. What is taken to be of interest under these rubrics is very widely and liberally (sometimes, perhaps, rather too liberally for some people's tastes) interpreted by officers of the Society. This line is, in general, well supported by members through their attendances at day conferences and other events like the week-long Dundee conference organised by the society. In the compilation of the present collection, the same attitude has been adopted, so far as has been possible. People have been encouraged to say what they wish to say in the ways in which they wish to say it.

Some of the chapters may sometimes seem rather speculative. Speculation, though, often leads to advances (cf., for example, Beidelman 1973: 162 n. 45) and as such is not to be abjured. Questions, though, do arise: What is being speculated about? Is it worth taking time and expending effort over? Is it likely that the matters speculated about, and/or the findings to which they lead, could be demonstrated (or could at least be shown rigorously to be probable, or not so)?

Others of the chapters are able to rely to a greater extent upon materials that are more, so to say, immediate. Perhaps Mr Sharma writes about the most immediate data in the collection, for these form an integral part of his own life (as they suggest they should). He describes some of the contests that a person encounters as he or she seeks salvation. Mr Sharma, as befits a Brahmin, is a teacher. He explains simply and concisely how one best seeks deliverance and how one knows when one has attained it. Nothing is irrelevant to one's progression along the path one elects to follow, nor to winning the contests encountered as one moves along it; while salvation or deliverance affects one's whole life (and after-life).

Independently, Duff-Cooper shows that what Sharma has to say about deliverance is an ideal to be sought in Balinese ideology on the island of Lombok. He does so by a consideration of beetle-matches, the winning or losing of which, as indeed the winning or losing of any contest, should inspire neither elation nor dejection in a winner or a loser but a more balanced state of mind. Beetle-matches are shown to evince principles of order and more substantive things, pragmatically speaking, that are discernible in other aspects of Balinese ideology. These findings suggest again that Balinese life is a (sacred) totality or whole of which all the aspects, if any, are ritual — as indeed they are for the person seeking or who has attained salvation. Finally, the chapter considers the use of formal notions in sociological analysis and raises the matter of fuzzy logics.

In Hoskins's study about the mock mounted combat (*pasola*) that takes place periodically in Kodi, western Sumba, she describes and analyses ideas and practices associated with it. The description and analysis entails recording events in, and other aspects of, village life, myths about the combat and the swarming of sea worms associated with the combat, and accompanying activity in Kodi. She also provides a historical and theoretical consideration of the combat. One may disagree, at times, about what especially the latter contends, but the chapter is anyway a most interesting, finely drawn, and provocative study that contributes importantly to knowledge about Sumba.

From Sumba, the move is to northwestern China, where song-exchanges in which anyone can take part are held at what used to be known as temple or pilgrim fairs. Lowry describes the songs exchanged and the contexts of their exchanges. She then suggests ways in which singing and other festival activities may be related and in which song may serve an 'integrative function'.

Next Japan, and Hendry's description and analysis of children's games there: One of the findings of this admirably detailed and engagingly unpretentious contribution to the literature about life in Japan is that 'contest and competition should be aimed out of groups . . .' (pp. 90–1) so that harmony can be maintained among their members. This finding might perhaps have been linked to Hendry's earlier suggestion (1987: 205) that equality is a principle that orders Japanese life along with various others including something generally just called hierarchy (as though this concept were unproblematical); and perhaps to studies by other writers (e.g. Duff-Cooper 1988) that suggest that equality, *contra* received wisdom, is indeed to be taken account of in the interpretation and understanding of Japanese (and indeed Balinese) life. But that it is not does not in any measure detract from the interest of Hendry's study nor from the importance of her findings.

Baquedano's paper concerns Mesoamerica, and particularly the ballgame that was played in the region from the Preclassical period (*c.*1200 BC) until the Spanish conquest in 1519. Baquedano's consideration of the various evidences available and of the relevant literature enables her to suggest that aspects of the ballgame have spread through the area, while others have remained the same throughout the area as other aspects of the game were modified. Also, she suggests that all the evidence leads to the conclusion that the ballgame, the movement of the stars, agricultural fertility, and death are all intimately connected. Interestingly, as a secular form of wrestling (*sumo*), once a

religious rite, is now popular in Japan (and elsewhere), so a
secular version of the ballgame is played today in Sinaloa in
Mexico.

Kehoe, naturally, writes about North America, specifically
about the Blackfoot, a confederacy of bands occupying the
Northwestern American Plains, now southern Alberta, Canada,
and adjacent north-central Montana. In the course of employing
historical and more recent materials for describing Blackfoot
contests of Power, Kehoe explicates concepts, relates myths, and
(like Baquedano in a sense) suggests how Blackfoot symbolism
may be just one transformation that a symbolic theme can take
within a cultural area. It is of note that this suggestion does not
allow linguistic and historical differences to impede it. Quite
properly, Kehoe is concerned with forms, and does not permit her
comparative endeavour to be limited by criteria of resemblance
that are not significant to that endeavour.

Davies's subject is Etruscan and Roman funerary art, and the
places of contests, and their various significances, in that art. She
considers their relationships with other aspects of life contemporary and antecedent to it, and, importantly, is able to
describe the polythetic characters that attach to the embellish-
ments of the artefacts considered and the meanings attributable to
them.

Ó Crualaoich is also interested in death, in the form that it used
to take in an Irish 'merry wake'. Ó Crualaoich shows that all the
various activities that took place then which non-participants
might simply adjudge bawdy or even licentious and wild are
readily explicable by reference one to another and to other series
of events elsewhere that take a similar form.

In focusing on European tradition, Lyle's essay is a fitting
conclusion to the collection. It returns to Scotland, where the
collection began. Her study may be viewed as a contribution to her
continuing concern to articulate the concepts that together might
define an archaic form of life, that is one that underlies or precedes
all other such forms. This concern has a most respectable past, of
course, having deeply affected the work of such scholars as
Durkheim and Lang, Wake and Thomas, for instance. That Lyle is
concerned with matters that in some quarters are rather out of
fashion is all to the good, for such work can be a powerful stimulus
to others not directly concerned with them (cf., e.g. Duff-Cooper
1986: 199–201). In her study, Lyle partly relies on Frazer for the
definition of her area of study and for comparative data to set
beside her more recent Scottish and other materials. Lyle's
ambition is here furthered by a consideration of winning and losing

in seasonal contests. Her study does indeed 'begin to sort out the bewildering tangle of seasonal traditions in Europe' and incidentally shows, too, the heuristic value of what Lyle calls a 'correspondence system'.

All the chapters implicate change in the obvious general sense that all are concerned with actions and relationships. But they also implicate change more particularly. Thus, some (those of Baquedano, Kehoe, Davies, ÓCrualaoich, and Lyle) might be called 'historical'. But these chapters are of two kinds: one (Lyle) has her eyes fixed ultimately on archaic society; the others consider 'lost' or 'obscure' traditions. Still, the questions all these chapters address concern what the data mean as and of themselves, or what they say about the forms of life considered. The other chapters (those of Sharma, Duff-Cooper, Hoskins, Lowry, and Hendry) describe ways through which change occurs, and the states that variously combine in different forms of life to constitute change itself, in process. Whether the topics each chapter addresses are important and interesting is a matter for each reader to decide for him- or herself. Naturally it is hoped that they are judged to be important and interesting. Can the findings of each chapter be demonstrated, or be shown through rigorous argument to be probable, or not so? Others, again, will themselves decide, but it is suggested that the findings of all the chapters could, in principle, be demonstrated or could be shown to be probable or not, as the case may be, because they are internally consistent, can take account of counter-examples while colligating all the known facts, and do not fly in the face of experience.

All the chapters, of course, consider contest. 'Contest', in line with Traditional Cosmology Society convention, has been widely interpreted by the contributors to this book (as it was by contributors to the conference whose papers are not included here for one reason or another). One striking and encouraging feature of all the chapters is that none is exercised by the classification of contests as, for instance, 'vicarious' (Ong 1981: 91 ff.) and other. In place of such a sterile exercise, the chapters together reveal how the ordinary English word 'contest' is an 'odd-job' word (e.g. Wittgenstein 1958: 43–4). As such it has no analytical value. But none the less it directs attention to a variety of social phenomena that have a family resemblance one to another and which (the studies collected here also show) can be approached and explicated in different ways, all of which are enlightening and cogent.

Knowledge, in a phrase of Mark Hobart (1985: 50), is built up from a plurality of perspectives. Less pragmatically, Needham argues (1983: Chapter 2) that considering material from as many points of view as he or she can is a constant and unevadable necessity for the analyst of social forms, but one that is a provocative incitement to the analytical imagination. The chapters in the present *Yearbook* provide such a range of perspectives. Interpretation is another such necessity (Duff-Cooper in press). But this exigency is liberating — as the Traditional Cosmology Society affords a welcome liberation from the confines of the presses, concerns, and distractions of particular disciplines and their leading ideas, as also from their proponents or those persons who seek to be their proponents. The studies that comprise this *Yearbook* also provide such a liberation.

To conclude, best thanks are due to Emily Lyle for the invitation to co-ordinate the present volume and to the contributors to it for making the task an enjoyable and rewarding one.

REFERENCES

Beidelman, T. O. (1973). Kaguru Symbolic Classification. In *Right & Left: Essays on Dual Symbolic Classification*, ed. Rodney Needham, pp. 128–66. Chicago and London: University of Chicago Press.

Duff-Cooper, Andrew (1986). Andrew Lang: Aspects of his Work in Relation to Current Social Anthropology. *Folklore* 97/2, 186–205.

—— (1988). *Aspects of Japanese 'Exchange': A Conspectus.* Occasional Paper 64. Free University, Berlin: East Asian Institute.

—— (in press). The Formation of Balinese Ideology in Western Lombok: Inspection, Change of Aspect, Assessment. *Sociologus*.

Hendry, Joy (1987). *Understanding Japanese Society.* London: Routledge.

Hobart, Mark (1985). Texte est un con. In *Contexts and Levels: Anthropological Essays on Hierarchy*, eds R. H. Barnes, Daniel de Coppet, and R. J. Parkin, pp. 33–53. Oxford: JASO.

Lyle, Emily (1990). *Archaic Cosmos: Polarity, Space and Time.* Edinburgh: Polygon.

Needham, Rodney (1983). *Against the Tranquility of Axioms.* Berkeley, Los Angeles, and London: University of California Press.

Ong, Walter J. (1981). *Fighting for Life: Contest, Sexuality, and Consciousness.* Ithaca: Cornell University Press.

Wittgenstein, Ludwig (1958). *The Blue and Brown Books.* Oxford: Basil Blackwell.

D. L. SHARMA

Different Paths to the Hindu Concept of Salvation

At first sight it might seem as though no contest is involved in seeking and attaining what Hindus understand as salvation (*moksha*); but the greatest contest in the life of men and women, in that of *all* men and women, is the struggle for liberation, salvation, or deliverance, which Hindu religious texts call *moksha* or *nirvana*. The Shastras insist that the most important and therefore the most worthwhile contest is for the attainment of this state of *moksha*. In life people on many occasions get involved in many kinds of contests, which may be in games, or sports, for the hand of a bride (cf. Duff-Cooper in this volume), for possessions, wealth, land and property, or for leadership, perhaps kingship, or ownership, or for religious authority. And most often these contests are played out on battlefields. All these are contests for worldly gains and material advantages. There is, however, another kind of contest in which people find themselves continuously and constantly engaged. This contest is between good and evil, between succumbing to temptations and overcoming them, or, if one likes to express it so, between going to heaven or to hell.[1] One may escape all kinds of contests, but this contest between good and evil is unavoidable. Other contests may be fought at the will and command of others, but fighting the contest for *moksha* is inherent in all of us. Benefit of victory or the loss of defeat in all other contests may go to other people, but in this contest one's self is always either the winner or the loser. Winning the contest for *nirvana* is positively for the benefit of one's self. This is the greatest contest and, therefore, if indeed it is a contest, the highest award accrues to the winner.[2]

The corpus of Hindu religious texts is huge, running to over half a million verses. Hinduism, of course, is an ancient religion, perhaps the oldest among leading world religions, and great saints, sages, and scholars have added to its literature over many

centuries. One need not, therefore, be surprised at the large size
of the Hindu religious literature. It is huge even when the last text,
Bhagwat Purana (accepted as, given the status of, and admitted
into, the category of Shastras), is not less than 5000 years old.
During the past fifty centuries, no text has been accorded the
status accorded to the Shastras, although many highly respected,
widely read, and indeed most valuable books were produced by
Buddhist scholars in and around 100 BC, and by authors like
Shankaracharya in the seventh century, followed by Ramanujach-
arya, Vallabhacharya, Madhusudan Saraswati, and many others.

All these texts, and the thousands of verses of Shastras, are in
complete agreement as regards (*a*) the meaning of *moksha*; (*b*) its
desirability as the ultimate objective; and (*c*) the conditions
required to be fulfilled to achieve the state. There are, though,
differences about how these conditions can be fulfilled. Broadly
speaking, there are three main schools of thought, or rather three
different spiritual paths recommended by these schools, by which
moksha can be attained. These are: *karma marg* (that is, the path
of actions and deeds); *bhakti marg* (the path of devotion and
surrender); and *gnyana marg* (the path of knowledge and
understanding).[3] Each of the schools claims that *moksha* can most
easily and quickly be reached by the path advocated by it. Before
discussing the claims of the supporters of each path, I should like
to consider briefly the three aspects of *moksha* upon which these
schools agree, namely, what *moksha* is, why it is necessary and
desirable, and who can attain it.

All the Hindu texts are agreed as regards the definition of
moksha, *mukti*, *nirvana*, liberation, salvation, or deliverance. The
agreed definition consists of two parts, one positive and the other
negative, but both are simple and straightforward. Thus, in his
commentary on *Skanda Upanishad*, Swami Krishnanandji says:
'Describing moksha, all Upanishads and Puranas say that moksha
is niratishaya suhk prapti and atyantic dukhak nivriti' (Krishnanandji 1980: 13–14), that is, attainment of a state of never-ending
and unalloyed happiness and complete and everlasting absence of
unhappiness.[4] Here by 'happiness' is not meant enjoyment,
pleasure, joy, hilarity, and merry-making, but a state of complete
calm, tranquillity, undisturbed equilibrium, absolute lack of any
want or desire, pressure or tension and a state of perfect bliss. All
are agreed on the meaning of this concept and this, in effect,
describes a state when nothing is left to be done, there is no desire
to do anything, no strivings, no trying, not ever wanting or
desiring; it could also refer to a state of nothingness. This may
sound somewhat anti-social, inhuman, even unnatural, and perhaps

even impossible. Shastras, however, claim that it is not only possible to attain such a state but they cite examples and life stories of quite a few persons — individuals such as Kapil, Yagyavalka, Rishabhdeva, Janak, Bhishma, Buddha, and others — who actually reached that ideal state and lived in it for ever after.[5] Such a state is perhaps unnatural to ordinary mortals but, according to Shastras, getting out of the rigmarole of nature, away from the laws of nature, is the objective of *moksha*. It no doubt seems inhuman and anti-social but, again, according to Shastras, humane considerations and social activities are bondages, an emotional involvement which ties a person to the miseries, difficulties, failures, and disappointments of worldly life. This involvement should be abjured, got rid of, and moved away from. In verses 26–9 of Part 1 of Chapter 1 of *Kath Upanishad*, Nachiketa, a young but devout seeker after spiritual knowledge, tells his teacher Yama: 'Sir, all the physical pleasures and worldly attainments, immense wealth and long life, diminish spiritual potential, so please keep all the young women, possessions and enjoyments, away from me, but teach me the most valuable truth about the *atma*, that which survives even after death.'

This raises the next question: Why is *moksha* a desirable objective? Why should one want, seek, and try to attain *moksha*, particularly when it seems to be anti-social, inhuman, unnatural, and perhaps nearly impossible to achieve? Shastras offer a cogent, realistic, and worthwhile answer to these questions, an answer to which the three different schools just mentioned subscribe. That answer of Shastras is that it is *moksha*, liberation, or bliss that all of us are seeking all the time, every moment of our life, without being aware of it. It may sound strange, if not absurd, to say that we are seeking *moksha* all the time, but Shastras assert that this is indeed so. If the meaning or definition of *moksha* is happiness, then it cannot be denied that we, all human beings, and maybe animals too, are perpetually seeking to achieve *moksha*. If this state is fulfilment of all our desires, a state of 'desirelessness' or 'wantlessness', then certainly all of us are striving for it all the time.[6]

One rushes through the morning routine, washing, taking a bath, dressing, having breakfast and getting to work, with the object of overcoming the tensions that force one to do all these things. One works hard to earn wealth, obtain recognition, attain position, acquire status, or gain respect in the hope of meeting one's needs, satisfying one's wants, or boosting up one's ego, because this is the cause of one's unease. One takes up an activity, any activity, with the sole fundamental purpose that when it is over one will be at peace. It is peace, ease, quiet, poise, happiness, liberation, or *moksha* that, Shastras aver, all of us are seeking and striving for. All our activities, from such simple tasks as the tieing of a shoe-lace to complex ones like mounting an armed attack on an enemy; from the simple task of buying a loaf of bread to such a complex one as equipping a factory for the manufacture of cars; from running a small family kitchen to administering a superstate government — all these tasks and the activities associated and involved with them are directed towards silencing our mental agitation so that we can have peace; towards fulfilling our desire so that we may reach a state of rest; towards relieving ourselves of inner tension so that we can be happy; or towards satisfying our ambitions so that we may overcome our misery. All these are attempts at reaching a state of mind of unalloyed happiness and getting rid of unhappiness. People are constantly engaged in this contest against pressure, tension, and unease. Shastras, however, say that this is a worthless pursuit, since none of the worldly achievements brings real peace, and this for two reasons. One is that these tasks and activities are all transitory and impermanent; the other is that none of them is unaccompanied by some other kind of misery and distress. They do not, therefore, take an individual any nearer to *moksha*. It is, therefore, a life-long contest.[7]

So Shastras claim that the only real contest that all human beings are faced with is to conquer the pressure of their desires and to obtain peace or bliss. Our opponent in this contest is the inner self, consisting, among much else, of desires, temptations, aspirations, and false hopes. We are misled, drawn towards seeking physical pleasures to be had from food and drink and sex; greed urges us to seek out wealth and possessions; many desire name and fame in society and entertain aspirations for the betterment of family and friends. These are the enemies against which one has to fight. In this contest generally people do not realise that indulging in pleasures ruins their health, and that even a great deal of wealth often does not satisfy a person. Acquiring the highest post in the land may not bring happiness. True happiness, release, and bliss is to be found only in renunciation and in giving up the desire for worldly possessions. One of the definitions of our life, one given in the *Shrimad Bhagwadgita* (9.33), perhaps the most prestigious of Hindu texts, is that it is 'anityam asukham', impermanent and full of misery. The next line of the same verse says, 'Therefore, give it [worldly achievement] up and become devoted to Me [God].' This is the main message of Hinduism and of many other religions too.[8]

The desire for the attainment of *moksha* is present in all of us, but it is being contested all the time by evil intentions, undesirable pursuits and sinful deeds. All our worldly achievements, if they are not outcomes of sinful and evil intentions even then, are worthless for spiritual advancement, as they remain behind and are lost when death takes us from this world. We are, therefore, losing the battle, getting nowhere near our goal, but find ourselves further away, and certainly far away, from it. How can one get what one desires most and desires most insistently, constantly, and continuously? Here, again, Shastras come to our aid and say (and repeat more than once) that fulfilling three conditions amounts to *moksha*. These three conditions are: all the knots of desires in one's heart must be untied; all doubts of one's mind must be removed; and all the *karmas* of one's past lives must be neutralised. Fulfilment of these three conditions enables one to win the contest and to reach *moksha* or to become one with God (meet the Father).[9] The three schools of Hinduism to which we have already alluded are all agreed on these three requirements for attaining *moksha*. The *Mundak Upanishad* (2.II.8) says: 'Breaking [all] bonds [of desires] from one's heart, removing all doubts [from one's mind], and exterminating all karmas [consequences of past deeds] one unites with God.'

Followers of the three schools also agree that renunciation — that is complete detachment from all worldly possessions, property, cash, home, status, and position, severing all connections with members of one's family, one's spouse, children, and relations, complete indifference to one's body, health, looks and well-being, and indeed cutting one's mind off, silencing one's intellect, and attaching oneself to one's own soul or spirit (*atma*) — is essential to the achievement of *moksha*. They also agree that this is possible in this life, here and now. In fact, they accept what Shastras say, that while one is in deep, dreamless sleep one does indeed become detached not only from one's surroundings but also from one's own mind and intellect. During those few moments of deep, dreamless sleep one nearly attains liberation, but not quite. One does not quite attain this marvellous state: first, because one is not conscious of the state that one is in; and second, because one soon returns to all one's usual worldly involvements and ties.

The agreement between the schools, however, ends here because they advocate different paths to salvation. Let us first consider the *karma marg*. This school says that *moksha* can undoubtedly be attained through renunciation. True renunciation, however, is an impossibility because the mind and the body go on functioning, operating, working, and acting so long as one lives.

No-one can exist for even a second without doing something (*Na hi kashchit kshan mapi tishtati akarma krit*).[10] One cannot detach oneself from doing something or other, so the only way is to go on doing, acting, performing, and functioning, but with complete detachment from the results of what one does, performs, and functions, with stony indifference to consequences, desiring nothing, wanting nothing, wishing nothing, hoping for nothing, and accepting whatever befalls one. The *karma marg* advocates that one should go on acting but give up all care and concern for the consequences. If one goes through life with such detachment, one will fulfil the three requirements for attaining *moksha*, and will certainly attain liberation. Verse 14 of Part 3 of Chapter 2 of *Kath Upanishad* declares: 'When one gives up all aspirations [and desires] that crowd into the heart [and mind] of man, then he who is [otherwise] mortal becomes immortal and certainly attains God-realisation.'

Those who advocate the second path, *bhakti marg*, base their views on the ground that the easiest, most pleasant, and most effective way of fulfilling the three conditions required for *moksha*, outlined earlier, is by surrender to God.[11] Accepting the theory of Shastras that God is everywhere, in everyone and supreme, one should offer all that one has, all that one does, and all that one receives, to God. Place everything at the feet of God and He will cut asunder the knots of one's heart, remove all doubts from one's mind, destroy all the effects of one's past deeds, and lead one to *moksha*. In the *Bhagwadgita* (12.6–7), Krishna gives a solemn undertaking to Arjuna: 'Those who surrender their hearts to Me, submit all their deeds to Me, completely depend upon Me, and are single-mindedly devoted to Me, are pulled out by Me, without any loss of time, from this ocean-like universe full of mortality and misery.'

Followers of the third school and path, the *gnyana marg*, say that it is true knowledge — complete understanding of reality, clearly knowing what the true nature of the universe is, what the individual self consists of, and what God is, and the relationship between these three — that will fulfil the three conditions and lead one to salvation. Once one realises that God is in everything, that He is functioning in one's mind, creating desires, controlling attachment as well as detachment, one will be left with no doubt or doubts, and the effects of one's past actions will be dissolved. According to *gnyana marg* theory, knowledge controls actions. In a desert one aims for, rushes for, a mirage so long as one thinks it is a pool of water and does not realise that it is a mirage. Once one comes to know that there is no water there, this knowledge will remove the desire and the wish to seek water in the sand. One gets frightened when one mistakes a piece of rope for a snake. Once one acquires the true knowledge that it is not a snake but a piece of rope, one's feelings, thoughts, and actions are altered as a result. A verse of *Chhandogya Upanishad* declares (8.VII.3): 'He [who has acquired true spiritual knowledge] succeeds in obtaining [all the pleasures that exist in] all the regions [of the universe, from earth to heaven] and he fulfils all his desires.' Once one is in possession of the knowledge of the truth, or reality, that no worldly achievement can bring one happiness, one will have no desire, no will, no determination to strive or work for, nor aspire to, any worldly achievements. Knowledge, or true knowledge itself, will do the trick and fulfil all the three conditions necessary and sufficient for *moksha*.[12]

In these ways each of the three schools of thought claims that its path is the best, the right, and, indeed, the only sure way to attain salvation. In the end as one comes closer to this blissful state, the three paths merge into each other and become one. Performing actions with complete detachment, that is with complete indifference to the results of one's actions, which is *karma marg*, is the same as complete surrender to God and placing everything at His feet, which is *bhakti marg*, and both these paths are equivalent to a realisation of the truth that nothing in this world really matters and then losing the will, desire, and determination to do anything at all (*gnyana marg*).

The three paths merge, or combine, into one for the simple reason that it would be hard to detach oneself from the consequences of one's actions or become completely indifferent to the results of one's deeds unless one were convinced that one's activities are all as futile as beating the surface of a mass of water with a stick. There cannot be complete surrender to God either without the clear realisation that hoping to achieve anything of abiding value is akin to hoping that one can get at the moon by diving into a river after its shadow. Similarly, there cannot be complete withdrawal from any interest in the world around without the clear realisation that all one's efforts in life are as futile and as wasteful as sitting on a river-bank filling one's cupped palms with water and then throwing it back into the river. None of these three paths can be followed without clear understanding and knowledge of the true nature and purpose of the universe.

For the attainment of such an understanding, Shastras prescribe three steps — *shrotavya, mantavya*, and *nidhisasitavya*.[13] The first refers to coming to know the truth by hearing about it from someone or reading about it somewhere. The second refers to the close examination of the truth that one has come to know. Finally, *nidhisasitavya* refers to the way one is convinced by, and comes to accept, the truth that the entire cosmos consists of one element, God; that all activities in it are the result of one force, God; and that there is no purpose, no end, and no objective for people in this world except to reach this truth and to be rid of illusions. This is the real contest in which every person, man and woman, is engaged. Winning the contest between the higher, nobler self and the lower, baser aspect of one's self offers the highest prize, *moksha*. Many in the past have gained this prize and, Shastras claim, it is within the capacity of each person to win the prize for him- or herself.[14] Thus, Krishna, in the *Bhagwadgita* (15.20), says: 'One who is able to understand and hold on to the deep truths that I have pronounced is left with nothing else at all to accomplish.'

NOTES

1 The view that the contest between the forces of good and evil in the hearts and minds of people is the hardest has repeatedly been emphasised in Hindu religious texts. The *Shrimad Bhagwadgita* (6.34) expresses this view in these words: 'It [the mind] is obstinate and powerful, ever fluctuating, capable of thoroughly churning the heart, and controlling it [from rushing towards evil] is as impossible as stopping the wind from blowing.'

2 The fact that the attainment of *moksha* is the highest goal or prize is pointed out in the *Kath Upanisad* (1.III.11), for instance, where the verse ends by saying, 'sa kasta sa para gatih', that is, the last limit and the final goal.

3 According to the *Kenopanishad* (IV.8), there are three bases upon which attainment of *moksha* can be founded: performance of (good) actions; engaging oneself in worship of and devotion to God with complete control over the senses; and acquiring knowledge (through the study of Vedas).

4 This definition of *moksha* appears in many Hindu texts. See, e.g., Krishnanandji (1980: 92).

5 Life stories of many who have attained *moksha* appear in different Hindu texts. That of Janak is mentioned in the *Shrimad Bhagwadgita* (3.20): Janak and others are said to have attained the final objective ('sanshidhim asthita Janakadaya').

6 The *Shrimad Bhagwadgita* (16.12–16) describes the futile attempts of all of us to achieve contentment through accumulation of wealth, by means fair or foul; through seeking physical pleasures; aspiring to more; defeating adversaries; becoming powerful, successful, happy, and better than others. These verses end by saying, 'patanti

narake asuchan', misguided by vagaries of mind and bound by ties of attachment such people get dragged down into the cesspool of hell.

7 In verse 7 of one of his seventh-century works, *Sar Tatva*, reprinted in 1927 by the Benares Hindu University, Shankaracharya says that even after he is old and has to walk with the help of a stick, his hair has become grey, he has lost all his teeth, his face has become pale, and his body disabled, yet 'na munchati asha', a man does not give up hope of finding happiness in worldly successes.

8 The *Shrimad Bhagwadgita* (9.33) stipulates 'asukham anityam imam lokam', the world is [full] of miseries and is impermanent; therefore, having come into it, worship God.

9 These three requirements for uniting with God are mentioned, for instance, in the *Mundak Upanishad* (2.8).

10 On people not being able to give up doing something or other even for a moment, see the *Shrimad Bhagwadgita* (3.5).

11 In verse 27 of Chapter 9 of the *Shrimad Bhagwadgita*, God directs, 'Whatever [good or evil] you do, whatever you eat, whatever sacrifices you make, whatever you give in charity, and whatever penance you perform, surrender [and submit] all of them to Me.'

12 Those who advocate the path of knowledge quite often quote: 'hrite gnyanat na mukti', no liberation without knowledge, 'na anya pantha vidyate ayanaya', there is no other way possible (cf. Krishnanandji 1980: 104).

13 For a detailed discussion of these three steps in attaining true spiritual knowledge, see Krishnanandji (1980: 271–5).

14 The last verse of *Kath Upanishad* (2.III.18) promises that just as Nachiketa had reached God, 'anya api', anyone else also, who follows the prescribed path can likewise reach God.

REFERENCE

Krishnanandji, Swami (1980). *Discourses*. Ahmedabad: Sanskrit Maha Vidyalaya.

ANDREW DUFF-COOPER

Balinese Contests on Lombok: Some Remarks[1]

I

Among the Balinese of Lombok, there are various contests occurring more or less sporadically. Among such contests (which, compared with other such events, may seem informal but which like every other kind of social action are conventionalised) are such situations as, for instance: when two men wish to take one woman (or two women are attracted to one man); or when one Pedanda ('high priest') tries to use his mystical abilities to the discomfiture of another; or when a practitioner of white, right-handed magic attempts to counteract the effects of the work of a witch, practitioner of black, left-handed magic.

More formal contests also take place between individuals who may or may not belong to the same local descent group or between groups of people from the same or more usually from different local descent groups. Examples of such 'individual' contests are cock-fighting, card games (especially *cuki*), fighting with staves or swords and shields, and beetle-matching. Examples of the latter, 'group', contests are mock-battles, like the one held at the important Balinese temple of Lingsar (see Duff-Cooper 1988a), where combatants hurl rice-cakes at one another which farmers participating in the spree may later collect and use as fertiliser in their rice-fields, and battles or fights which are far from mock between people (men) from one or more Balinese village (*kaklianan*) and, as once happened when I was living in Lombok, Muslims from other villages nearby.

This brief study focuses on one form of contests — beetle-matches — but I suggest that what analysis reveals about these contests also holds for all Balinese contests, by implication. Whether this surmise is well-founded is an empirical question which will not be addressed further here.

II

Young (i.e. adolescent and usually unmarried) men often keep a beetle (a stag beetle, I think) called *jankrik*, which they have found or come across, in a piece of bamboo about six inches long and one and a half inches in diameter which is hollowed out, its length slatted to give a view of, and air to, the inside, and is plugged at one end. From time to time, often when sitting around with friends in the shade of a compound to escape the midday sun, a beetle owner will challenge another to a match. (I do not know if male beetles are matched against males, females against females, or indeed whether there is any preference for male or female beetles as combatants in these matches.) A beetle is introduced into the container of the challenged insect. The insect that is by general consent of those present determined to have retreated in face of the other loses.

Unfortunately, there are a number of other matters connected with these matches about which I regret I cannot be very informative. The first is betting. My recollection of this aspect of these contests is that betting on beetles matched against one another is fairly simple compared with the betting at cock-fights, for instance, which can involve fairly complex arrangements between two or more people. In beetle-matches, the two sides to a bet lay money on the likely outcome of the match. The side that correctly predicts the outcome takes the money laid by the other side and takes back its stake. If a draw is declared, and unless one side has predicted such a turnout, the bets may be withdrawn or laid on a re-match. A contract is involved, therefore, which is not unexpected given the ubiquity of this institution (but see Duff-Cooper 1990: 141).

Next is the matter of who challenges another to a match. There are no formal factors — such as relative age and/or estate (*bangsa* or *varna* of which there are four) — involved here, and 'sex' is certainly not, for young women may be present at matches but in my experience do not keep beetles to match against others. Indeed, a challenge may be issued simply because the challenger is bored and wishes to liven things up. A challenge instantly brings the proceedings to life if the challenge is met and betting begins, or someone who is reluctant to take up the challenge is urged by those present not to be a killjoy but to meet the challenge. Here is evinced a characteristic of all Balinese events that have come to be called rituals: the quality of liveliness (*ramé*) which often takes the form of many people and much activity and various sounds (e.g. music, speech, chopping, pounding, plus the other more usual sounds of life in the village such as the crowing of cocks, the grunting of pigs, the wailing of children, and the screaming of crickets) and which is taken to mark a contrast with the ordinariness or tedium of the normal daily round. This is what marks contests and other rituals (*yadnya*, 'sacrifices') as such: the view that associates rituals with the sacred falls on two grounds in the context of Balinese life. First, there is no distinction 'sacred/ profane' in Balinese ideology; second, all aspects of Balinese life, if any, are rituals. These sentences, incidentally, mean what they say. There is no distinction 'sacred/profane' in Balinese ideology but one which (as it were) takes its place in the scheme of Balinese things, *niskala/sakala*: the timeless, essential, invisible, and the material and visible and in time. All that exists belongs to one of these classes, or sometimes to one and sometimes to the other. All that exists, further, is a reflection of, and is pervaded by, Ida Sang Hyang Vidhi, the high or highest god of the Balinese, Order, What orders to borrow Mark Hobart's (1986) phrase. Balinese forms of life (*dharma*) are equally aspects of Vidhi and as such may be conceived as being all ritual.

Finally, I should have liked to be able to say something definite about the criteria that make a good bettle, such that someone chooses to cage it and then to pin himself to that beetle in matches or to bet on it. Perhaps whether a beetle is a good one is established simply by the outcomes of matches, that is, there are no criteria attaching to 'good' beetles as there are to Western race-horses or pedigree cats and dogs, for example. Someone who is confident of his position with the gods — one who has pleased them is likely to be given the gift (*pica*) of victory, here as elsewhere, one who has displeased them, defeat — challenges a well-established beetle and gains its prestige if his beetle wins. If it loses, little is lost. Someone whose beetle consistently or often wins is likely, of course, also to have a greater following among punters than someone who is known to have had few or no winners; though one is not betting on the owner so much as upon his beetles, his followers know he has an eye for a winner.

It is clear though even from the sparse material presented that beetle-matches and their outcomes, by linking an individual and the people who support him through laying bets that favour him above his opponent and supporters establish whether relationships in this context are symmetrical or asymmetrical. All such relationships are gauged by reference to a point which takes different forms in different contexts. Here the point may be said to be the gods in the contestants' compound or village temple (*sanggah kemulan* or *pamaksan*). These gods decide, as it were, whether one is to win or to lose and, hence, whether in this context one is related symmetrically or asymmetrically to one's opponent. If one wins, one's relationship to one's opponent is asymmetrical and one is superior: one is favoured by the gods and so are one's supporters; one's opponent and his supporters are not and are inferior in this contest. Balinese beetle-matches (like cock-fights, card games, and all situations that result in a winner and a loser or losers or a tie) are a way of establishing order in the non-functional sense of the relation that obtains between two entities which may each, of course, be either individual or several.

The relation obtaining between two individuals, A and B, of whom A sometimes wins and sometimes loses and B generally wins but also loses sometimes, is more symmetrical or less asymmetrical than that between B and C, for instance, where C sometimes wins but generally loses. Here, allowing that A, B, and C all place more or less the same small amounts (perhaps as little as Rp.10) as stakes on a match, B is superior to A and C, but A is superior to C. This relation is non-transitive, of course, there being nothing necessary about the relations obtaining between A, B, and C.

Before this section is closed, it should be said that Gerhart Baumann has suggested[2] that he is not sure whether 'winning pure and simple establishes superiority. There must be "deserved" wins and "freak" wins? Wins credited to skill, and wins put down to chance or beginner's luck.' These, of course, are matters that need to be decided empirically, which I am not in a position to do now. So I cannot answer Baumann's point about deserved and freak wins, except by saying that my understanding is that all wins are deserved, in that they reflect the goodwill of the gods, and thus none is freak — an understanding naturally that leaves no room for skill, chance, or beginner's luck, all of which are certainly ideas that form a part of Western tradition but which cannot be assumed to be parts of other, exotic traditions.

III

At least three kinds of questions arise. The first concerns contests in general. Is it a useful analytical category? It certainly pinpoints interesting areas of social life and, in line with the Trantric precept which avers that all things, from Brahma to a blade of grass, should be our teachers, directs attention to institutions which, like beetle-matches, might otherwise be dismissed or at least overlooked as trivial, simply an adolescent diversion.

This raises the question, next, of the agonistic content of contests. Among the Balinese of Lombok, many contests are diversions about which people (sometimes the same people) are avid, whereas most Balinese go out of their way to avoid fighting. Balinese contests, therefore, can be seen as a class of events some of which are a form of entertainment, but others of which (matters of life and death) may have no diversionary content. This is not to say that diversionary contests, as it were, are less important than those that have no such component analytically speaking; nor that in certain circumstances — when one's ducks have been interfered with continually and unnecessarily, when the gods and the realm are taken to be threatened — fighting to the death is a bad thing: on the contrary. On the other hand, such fine beings as Pedanda and those who would emulate their purity and fineness do not attend events like cock-fights, where people gamble, drink red palm wine (*tuak*), and otherwise get 'hot', and inclined to anger and other unfine behaviour, words, and thoughts.

In all cases, though, the responses to winning or losing should be the same: a recognition of the fact that involves no element of celebration or despondency or their likes. Such extreme responses are unbecoming, and are echoed in various other responses that a fine Balinese, man or woman, cultivates: one should generally live a moderate life; one should neither enjoy nor dislike working, which includes holding and/or particpating in rites (*yadnya*), but one should just get on and do what in the situation one is supposed to do; one should be neither boastful nor diffident, vain nor wholly unconcerned about one's turnout; and one should not think too much nor be entirely unreflective.

Such responses are corollaries, though not necessary ones, of two related matters: first, that in the Balinese world people do not live. Life (*atman*, the soul, totally *dharma*, a form of Balinese life centred on one place, which may be made up of parts that are not physically contiguous, at a particular time) rather, lives itself out through people. And second, that life is ordered by certain principles, all of which are aspects of Vidhi, the high god just mentioned.

That is to say that joy and sorrow (and their likes) in response to such matters of fact are rather negative responses; more positive, and hence superior, is acceptance of them and a cultivation of the conditions for Balinese life to be able to live itself out as finely as possible.

Conventional or ideal responses to winning or losing are both an aspect of those conditions and of Balinese life at its finest. One aims, or should aim, for a response where symmetrical relations between, say, elation and despondency or dejection may obtain. This response itself evinces Balinese life at its finest: Vidhi is symmetry; asymmetry is also an aspect of Vidhi, but this mode of relation is less fine. Equally, these latter social facts are not matters for elation or despondency or dejection, but emerge through the cool consideration of less abstract matters (Durkheim's injunction to treat social facts as things comes to mind here).[3] The stance that allows the latter facts to emerge is equally that which a participant in the present case a contest of some kind best adopts. And the facts that these conventional or ideal responses that are evinced in other forms in other areas of Balinese beetle-matches (and I suggest other contests) order other areas of social life too (see Duff-Cooper 1987a, 1987b, 1988b, 1990), suggest that the analysis is one proper way (i.e. one that is faithful to Balinese ideology) in which contests can well be approached.

The relations just mentioned are: duality, symmetry, asymmetry, transitivity, reflexivity, analogy, homology, and reversal or inversion. Duality clearly pervades beetle-matches: the insects, the owners, the betting, the relationship of gods to the people involved in the matches, for instance, all evince duality. Symmetry, asymmetry, and transivity (in its non-transitive mode) have been alluded to above. Reflexivity is in evidence here as this mode of relation is inevitably in evidence in other areas of social life (see Duff-Cooper 1988c): a tie evinces reflexive relations (in the context of the match two contestants stand in the same relation to the contest in regard to one another as they do to themselves); a win or a match lost evince irreflexive relations: the two contestants cannot be both winner and loser at the same time. By virtue of these principles or order, one match is an analogue of any other. Homologies are therefore established among the participants in the matches. While reversal or inversion may, of course, be present in matches in which the same contestants compete against one another or when someone is sometimes a winner or a loser, it is subsidiary, as a principle of order, to those already expounded. Alternation (e.g. Duff-Cooper 1986) appears to be discernible only, so to say, by chance: it is not a formal property of Balinese contests in the way in which the other relations discerned are.

One slight puzzle might seem to remain before the final set of questions alluded to is addressed. Why, if symmetrical relations are finer than asymmetrical relations, should so many Balinese people take part in events that generally result in asymmetries?

To answer this question, one might resort to a putative universality of contests: all forms of life, it might be claimed, make room for such events, so it would be odd, and thus a matter for consideration, if the Balinese did *not* take part in such events; and one can imagine that non-hierarchical forms of life such as the Penan and the Chewong (see, e.g., Needham 1964; Howell 1985,

1989, esp. pp. 51–2) do not make such contests available as a form of social action. Certainly it is not a necessary part of any social grouping: it was explained to me once by Rodney Needham that working in a small social anthropological *équipe* did not, or need not, involve any kind of race; though then one is reminded of the fact that in bull races that were before (*sané dumun*) held in northern Bali — never, so far as I know, held on Lombok — the winning bull was not the one first across the finishing-line but the bull that crossed the line after running the most elegant race.

My view, at any rate, is that it is more convincing to argue that there is in fact no puzzle to be solved here. One does not, if a Balinese, properly exult in, nor feel depressed and cowed by, the asymmetry that is created among people through contests. One accepts it, as 'we' accept that asymmetry exists as a formal relation. Asymmetry is an aspect, though not a pervasive aspect, of Balinese life on Lombok, as it is of other forms of life, Balinese and others, and there is no more difficulty with it being evinced through contests than there is in it being evinced by other aspects of Balinese life. What matters is one's response to the mode of relation not the fact of its existence — though one may adopt a critical attitude to expressions of the relation and Balinese people often did when I was there: a venal, haughty, and proud Brahmana woman married to a Pedanda, for instance, was generally roundly condemned by those inferior to her because of birth and sometimes also because of their ages for her unfine (*nénten alus*) way of going on.

Baumann (see note 2), by the way, suggests that my critics might ask: Why evince asymmetry on contests, then, and not on data more widely known? There are at least two responses to this question. First, that the present study was written in its original form for a conference on contests (see note 1); second, that the data presented about beetle-matches evince asymmetry (and the other relations mentioned) is a very good example of what has been maintained in analyses of more mainstream topics,[4] namely that the relations discerned in those analyses are discernible in other seemingly quite different aspects of Balinese life; and as all that life is ritual, it makes no sense either by the tenets of Balinese ideology or of analysis to rule out certain data in advance as somehow unworthy of attention. As the Tantric precept urges, all things are, or should be, one's gurus.

The final set of questions, to which we can now move, is more theoretical. In the discussion above, reference is made to symmetrical and asymmetrical relations that are more or less so. Needham has suggested to me in a series of letters that earlier uses (e.g. Duff-Cooper 1985) of such formulations are fatally flawed because 'symmetrical' and 'asymmetrical' (it is claimed) are not predicates of degree but absolute predicates that describe relations that do not admit of further qualification. And while the mathematician Dénes Nagy[5] has written to me (in a letter dated 7 March 1989) that he likes the way the notions of degrees of symmetry and of asymmetry enrich these two concepts, he also writes that 'the symmetry–asymmetry pair represent only two extreme cases, and we may have various dissymetries. . . . (A) dissymetry describes the missing of some possible elements of the theoretically available symmetries of the system. . . .' I think that Nagy's views are useful, especially when it comes to the assessment or computation of the degree to which a particular relation is symmetrical or asymmetrical (see, e.g., Duff-Cooper 1987c). But Nagy's view that 'the symmetry–asymmetry pair represents only the two extremes . . .', if accepted, seems to commit one to agreement with the views of Needham just described. For various reasons, I am not yet prepared to accept Needham's view nor, if the implications drawn from his remarks are not wrong-headed, Nagy's.

Some of these reasons are as follows. Pragmatically, 'degrees of symmetry' and the more elaborate 'degrees of asymmetry' appear in the Balinese case to lead to results that are more enlightening than those that would result from the employment of 'symmetrical' and 'asymmetrical' as absolute predicates.

This suggestion may seem to beg the question raised by Needham. But I see no reason at this stage to be constrained by the work of logicians as I think Needham's argument is constrained (see, e.g, Needham 1983, 'Reversals'). After all, various logics compete for hegemony as that which, not to put too fine a point on it, is right; fuzzy logics (e.g. Zadeh 1975) have been constructed because traditional two-valued logic cannot take account of much, perhaps most, argumentation, so that the truth-tables of fuzzy logics incorporate a value 'neither true nor false'; and perhaps most impressive, Balinese ideology gives evidence of at least one relationship that gives the lie to the claim that the phrase 'is an ancestor of' cannot evince a relation that is reflexive because one cannot be oneself and be one's own ancestor at the same time (Duff-Cooper 1988b). And, of course, the notions seem to make good sense of the materials analysed.

Furthermore, since compiling the first version of these remarks, I have read Campbell's book (1989) about the people of the Amazon forests who call themselves Wayãpí. Here (Campbell 1989: 111–13) there is a reference to an essay by Heidel (1906) called 'Qualitative Change in Pre-Socratic Philosophy'. It is argued by Heidel (1906: 343, 346 n. 28) that for pre-Socratics such as Heraclitus, qualities were viewed as physical constituents present in a whole, i.e. that it contains qualities in a way that is purely physical. Conceiving of 'symmetry' and 'asymmetry' in this way allows the possibility of degrees that can be assimilated with other modes of relations when expressed in physical terms (e.g. through lights[6]) in a way that the abstract mode of thinking about these relations, and the forms they can take, may not permit.

This is a large and complicated topic, but it is one which I thought might usefully, and prudently, be mentioned in a study that aims to build on (and perhaps extend) the insights of Downs's study about 'Head-hunting in Indonesia' (1955), which I consider among the most interesting and stimulating pieces I have read connected with contests, and those of the remarkable body of work that Needham has done on the employment of formal concepts in the understanding of alien and exotic forms of life.

IV

To summarise in conclusion: Balinese beetle-matches, like other Balinese contests, are institutions that are substantively and formally variations upon themes and principles that are discernible in all kinds of other areas of Balinese life. Some are considered diversionary, some are not — being matters of life and death; while one may have a good time taking direct or indirect part in some matters of life and death (like procuring a wife or being taken by a man in marriage). Would it be going too far to suggest that, making the appropriate changes, these findings apply to the contests of all forms of life everywhere and at all times?

NOTES

1 The present study was written in May 1989 without the writer having access to his papers or to books, his own or others'.

2 In a letter dated 8 June 1989, which he has kindly allowed me to cite. Dr Baumann is the author of an acclaimed book (Baumann 1987) about the Miri of the Nuba Mountains in the Sudan, and currently lectures in Human Sciences at Brunel University.

3 Gerd Baumann (see note 2) comments here that he would would not bet on Durkheim's injunction being endorsed by colleagues, apparently because *Suicide* did not result in any social facts. I think, by contrast, that the notion of a social fact as something that we think we may know about a form of social life and treating them as things are both useful and

stimulating, though a demonstration of this must wait until a paper called 'Models and Modelling: Sculpting Balinese Ideology' (Duff-Cooper forthcoming) is presented publicly.

4 For references to some of these analyses, see the entries under the present writer's name in the Bibliography of Duff-Cooper (1988c).

5 A founding member and President of the International Society for the Interdisciplinary Study of Symmetry, Budapest.

6 The paper mentioned in note 3 above also goes into this matter in far closer, and practical, detail.

REFERENCES

Baumann, Gerd (1987). *National Integration and Local Integrity*. Oxford: Clarendon Press.

Campbell, Alan Tormaid (1989). *To Square with Genesis. Causal Statements and Shamanic Ideas in Wayãpí*. Edinburgh: Edinburgh University Press.

Downs, R. E. (1955). Head-hunting in Indonesia. *Bijdragen tot de Taal-, Land- en Volkenkunde* 111, 40–70.

Duff-Cooper, Andrew (1985). Duality in Aspects of a Balinese Form of Life in Western Lombok. *Cosmos* 1, 15–36.

—— (1986) Alternation and Other Modes of Periodicity from a Balinese Form of Life in Western Lombok. *Southeast Asian Studies* 24/2, 181–96.

—— (1987a). The 'Tone' of Balinese Ideology: Causal or Correlative? *Studies in Sociology, Psychology, and Education* 27, 109–20.

—— (1987b). The Balinese Rice-Planting Rite of *Nuasén* and the Magic Square of Three. *Cosmos*, 3, 41–57.

—— (1987c). Living with the Structure of a Balinese Form of Life in Western Lombok. *Philosophy* 84, 193–226.

—— (1988a). Balinese Pilgrimages: a Comparative Note. *Seitoku Kenkyu Kiyo* 21, 55–64.

—— (1988b). The Formation of Balinese Ideology in Western Lombok. *Philosophy* 86, 151–98.

—— (1988c). Reflexive Relationships in Aspects of the Ideology of the Balinese on Lombok. *Philosophy* 87, 219–55.

—— (1990). *Shapes and Images. Aspects of the Aesthetics of Balinese Rice-Growing, and Other Studies*. Denpasar, Bali: University of Volayana.

—— (forthcoming). Models and Modelling: Sculpting Balinese Ideology. *Journal of the Anthropological Society of Oxford*.

Heidel, W. A. (1906). Qualitative Change in Pre-Socratic

Philosophy. *Archiv fuer Geschichte der Philosophie* 19, 333–79.

Hobart, Mark (1986). A Peace in the Shape of a Durian, or The State of the Self in Bali. Paper delivered to the Leiden Workshop on the Balinese State, March 1986.

Howell, Signe (1985). *Society and Cosmos. Chewong of Peninsular Malaysia.* Singapore: Oxford University Press.

—— (1989). 'To be angry is not to be human, but to be fearful is': Chewong Concepts of Human Nature. In *Societies at Peace: Anthropological Perspectives*, eds Signe Howell and Roy Willis, pp. 45–59. London and New York: Routledge.

Needham, Rodney (1964). Blood, Thunder, and Mockery of Animals. *Sociologus* 14, 136–49.

—— (1983). *Against the Tranquility of Axioms.* Berkeley, Los Angeles, and London: University of California Press.

Zadeh, L. A. (1975). Fuzzy Logic and Approximate Reasoning. *Synthese* 30, 407–28.

JANET HOSKINS

Equal and Unequal Contests: Men, Horses, and Gods in Sumba's *Pasola*

I

In February of each year, thousands of horses and their riders throng to three different open fields along the western beaches near the ancestral villages of Kodi, Sumba, to participate in the *pasola*, a mounted ritual combat pitting members of the core villages of the region (Pola Kodi — 'the trunk of Kodi') against the more recently founded villages of the other side of the Wuku river and the embankment (Bali Hangali, Bangedo). For three or four hours at a stretch, riders holding blunt bamboo lances gallop out towards each other individually or in groups, throwing their lances to strike their opponents or their horses and knock them off. Then, they turn sharply to circle back to their own side to prepare another charge. Although rarely fatal, the combat is dangerous: both horses and riders often fall in the midst of the fray, and may be wounded and blinded by lances thrown. Those who have come to the combat after making the proper offerings and observing taboos are supposed to be protected by ancestral spirits. Others who have violated traditional custom may be exposed to super-natural sanctions, shedding their own blood as part of the sacrificial rites which precede the harvest.

The combat is held in conjunction with a calendrical rite, the festival of the sea worms (*nale*), which swarm along the western beaches once each year, usually in February and March, seven nights after a full moon. The tiny, multi-hued, nocturnally active Eunicid worms swim to the surface of the early morning tides to discard their genital parts after reproducing, performing a colour-ful dance, and providing the local inhabitants with a new and wondrous food from the sea. This natural display of vitality and fertility is greeted with its social counterpart in the *pasola* and its associated ritual. A human spectacle of bravery and daring, coupled with sacrifices and offerings that link the abundance of the sea to the abundance of the crops, symbolically binds the cornerstone of the traditional lunar calendar to the cycle of seasonal activities in the solar year. The date of the sea worm swarming is carefully determined by hereditary priests, in a complex calculation that has its roots in a mythical mandate.[1] Combining the attributes of sacred calendars and secular entertainment, the *pasola* is a religious ritual, the largest unifying ceremony of the 50 000 inhabitants of the Kodi district of the western part of the island, but also a huge public spectacle, a sporting event, and (local government officials hope) soon a tourist attraction.

In this piece, I will examine the *pasola* (1) within the ritual context of its performance; (2) in relation to mythic traditions of its origins; and (3) as it has been historically influenced and transformed. I argue that the *pasola* belongs to a class of ritual combats which are found throughout Indonesia and have been studied as evidence of dual organisations (Downs 1955), complementary opposition (Fox 1979), and sacrificial substitutions for warfare (Reid 1988). While each of these interpretations sheds some light on the dynamics and excitement of this spectacle, its place within the local political context also reveals another set of meanings, related to indigenous tensions between principles of equality and hierarchy. It is in terms of these two very different kinds of contests — one whose outcome is always predetermined; the other which constantly threatens to reverse existing inequalities in a more enduring fashion — that the contemporary performance must be understood.

II THE SUMBANESE AND THEIR HORSES

Sumba is an arid outlying island, fourth in the chain that extends east from Java and Bali towards Timor and New Guinea. Long a backwater which attracted little official attention, Sumba was known for several centuries for its production of fine *ikat* textiles and lively small horses exported along with slaves in an erratic and often illegal trade system (Needham 1983). It is at present the only Indonesian island where adherents of the traditional pagan system of ancestor worship (*marapu*) still form the majority of the population (Fox 1987), and it also has the most developed 'horse culture' in Southeast Asia.

The island's current human population of 400 000 cares for a roughly equal number of horses, as well as substantial herds of buffalo and cattle. Livestock circulate within the 'prestige economy of brideprice payments and feasting, where payments from wife-takers made in horses or buffalo must be reciprocated with countergifts of pigs and cloth. In the westernmost district of Kodi, subsistence is based on the cultivation of mixed gardens of rice, corn, beans, and various root crops. As pastoral agriculturalists, the people of Kodi define wealth through ownership of gold or animals, and give little social value to garden produce. Competitive feasts held to consecrate a cult house or ancestral village, to construct megalithic tombs, or to thank the spirits for earlier blessings are arranged in a series which culminates in the achievement of titles of renown.

The first set of these titles takes the form of 'horse and dog names': couplets ostensibly serving as names for a man's closest animal companions (such as 'the horse as fast as a bullet, the dog who barks ahead', *ndara pelor, bangga nggoko ulu*). In fact, these names indirectly describe the exploits of a particular man or a famous ancestor (here, qualities of speed and excitability), which are also used as honorific forms of address. All prominent Sumbanese men share a name with their favourite riding horse, and many will also 'pass on' the name to grandsons and other descendants after their death. The name is bestowed on a quality stallion in a special rite, involving the sacrifice of a chicken and recitation of a dedication. The ritual relationship thus established between the two is then understood to last throughout their lifetimes, and extend even into the afterworld.

A sample text of a prayer dedication indicates some of the imagery associated with the linking of their fates:

> Ta yo baka a ndara ole ura
> Ta yo baka a bangga ole ndewa
> Katu pa halako etu waiyo
> Tana ambu waingo wawi hyayo
> Tana ambu waingo ghai la mbyapa
> Timbu mburu mbanoho ela aronggu
> Helu mangico mangera la hadanggu
> Ambu wai kalu kingyoka
> A katondo nggu ura
> A halodi nggu lodo
> E byangoka yemi a ndewa ambu, ndewa nuhi
> Kahongga ndandi a mbale nda
> Kalete ndani a wongga nda
> Ambu halu taru puti
> Ambu paloho bei pa ana
> Enga ba mu ela ornona

> Let you be the horse friend of fate
> The dog companion of the spirit
> As I wander along the pathway of life

So there will be no pigs blocking the path
So there will be no logs lying on the trail
The heat will pass before me
The leafy shade will follow behind me
Nothing will come to
Strike me like the rain
Burn me like the sun
And you the spirits of ancestors and forefathers
Sit astride my shoulders
Ride across the back of my neck
To protect me in the moonlight
To shield the sow about to give birth
Providing for me along the way

Through these words, the destinies of horse and rider are linked, and they are committed to mutual guardianship and protection. The horse cannot be sold or given away once it has been dedicated to the spirits, and it must lead the funeral procession that will take the master's body to his stone tomb. Once the master is buried, the named horse must either be sacrificed beside the grave or given to his mother's brother, the 'source of his life', who bears the responsibilities for handling the body and receiving the dead man's personal effects.

The horse may carry on a name used by an ancestor, or be given a new name reflecting his master's biography. Sometimes the name describes the physical or psychological attributes of the horse (*ndara mburu manu*, 'the horse that gets up with the chickens', for an early waking and attentive horse; *ndara menahu*, 'ash-grey horse'; *ndara mete kabalyako*, 'horse black as thunder'; *ndara katupu*, 'short-tempered horse'), which the animal is believed to share with his master. At other times, it reminds others of experiences or attitudes which the master wants to remain in public memory. Examples of this kind of name are: *ndara honga dadi*, 'the horse who cried at birth', for a man who suffered many childhood illnesses; *ndara tanggu holo ole*, 'the horse who carries another's burdens', for a man who was falsely imprisoned; and *ndara marapu*, 'the horse of the ancestral spirits', for a man who refused to convert to Christianity.

A special divinatory technique can be used to 'match' a man and his 'name horse', by studying the patterns of swirls of hair. For a man, the 'soul's fate' (*ura ndewa*) is said to reside at the crown of the head, and specialists claim to be able to study a young boy's hair swirls and predict how many wives he will take, and which women will or will not prove suitable wives. Compatibility with a horse is evaluated in somewhat the same fashion. The swirls on the horse's neck and cheek should move in the same direction as those of their master, and have a similar location. The placement of additional swirls (i.e. on the withers for speed, on the back for endurance) provides clues to the animal's intelligence, disposition, character, and strength. Ideally, when the two sets of readings match, the man and his mount 'share the same soul and the same swirls' (*mera ndewa mono a urana*), and will be perfectly suited to each other.

The named horse extends his master's personal identity and serves as his symbolic double. The relationship is expressed not only in conventional respectful address (where the man is called with the name of his horse) but also in a series of metaphoric couplets which refer to a man's inner subjectivity through references to his horse. Thus, a man who suffers pangs of conscience is said to be 'the horse who steps on his shadow, the dog who barks at his reflection' (*ndara ndali magho, bangga nggoko ngingyo*). A person famous for his fierceness, virility, or eloquence is called a 'horse with an erect tail, a dog with a black tongue' (*ndara ndende kiku, bangga mete lama*) — reflecting also the Sumbanese custom of trimming the tailbones of stallions so they will stand upright and fly in the wind. The horse name is a symbolic vehicle for the transmission of reputation and fame across the generations. Serving as a mnemonic vehicle for particular exploits, it is said of a great man that 'his personal name was known, his horse name was renowned' (*na pa hada tamu touna, na pa ndende ngara njarana*).

Horse and dog names are given in conjunction, but it is only the horse name which is used for respectful reference. A Sumbanese man who is particularly fond of his horse may also depict it, painted or carved, on his tomb (where by custom the master himself may never be directly represented), and even bury it in a special chamber next to his own. One of the finest East Sumbanese textiles to travel to Kodi in a royal marriage eventually vanished from circulation as the funeral shroud of the former raja's favourite horse. The intense degree of identification between horse and rider also has the consequence that Sumbanese feel a repugnance to eat horse meat — although their close neighbours, the Savunese, relish it. When horses are sacrificed at funerals, none of the dead man's kin will take any portion of the meat, saying that to do so would be akin to 'eating the body of their own relatives'. Low-ranking outsiders, slaves, or Christian converts sometimes scavenge the remains, but horse meat is never served to guests or publicly distributed like pork or buffalo.

III THE PERFORMANCE OF THE *PASOLA*

In *pasola* performances, the speed, agility, and co-operation of horse and rider are put to the test in a ritualised combat which brings together male values of fierceness, aggression, and violent conquest, and female ones of deference, nurturance, and reproduction. Several thousand riders gather in the coastal villages to greet the sea worms addressed as a female deity (*inya nale*, 'Mother Sea Worms') and to 'entertain' her with this spectacle played out along the sea shore.

The biannual swarming of the *nale* is triggered by a set lunar phase and seasonal and tidal rhythms, allowing it to serve as an anchor for the traditional lunar calendar. *Nale* reproduce nocturnally in the hours before dawn, as they lie attached to the bottom of rocky reefs and coral banks. Once this sexual ballet is finished, they release long, many-coloured strings of genital products. In the early hours of the dawn, these posterior parts swim to the surface in strands of green, red, blue, or orange, where they are collected by the local people in bamboo sachets and buckets. Then they are brought back to the ancestral villages, where they are eaten as a condiment in the sacrificial meal which follows the first ritual combat. As they arrive, they must be greeted with the *pasola* — the spectacle specially mandated in Kodi mythology to entertain these unusual supernatural visitors.

Nale are associated with a renewal of vitality at the beginning of the new year, and also have a divinatory and propitiatory significance. The abundance of the sea worms augurs well for a successful rice harvest, and the letting of human blood in the *pasola* battle provides a compensation for the taking of a source of life and vitality from the sea. Kodi informants were always emphatic that the two events must be conjoined: it is prohibited to ride horses in *pasola* fashion, even for training, if the sea worms have not swarmed. It would also be extremely dangerous to wade into the tides to collect the worms if they were not properly greeted by the spectacle of the mounted combat.

The festivities are initiated by the Priest of the Sea Worms (Rato Nale), riding on a special named horse,[2] who circles the field at the beginning and closes off the action at the end. His concentration of sacred powers is supposed to afford protection to the riders on the field, preventing serious injury or death from the striking lances for those who respect the taboos. For those who defy them, violent collisions and falls may occur. In all the combats witnessed, minor injuries were frequent, and stampeding horses often crossed the lines into the enemy camp. The contest is a forum for the display of individual feats of daring horsemanship and skill, but it produces champions rather than winners. There are no final victors, and members of each side usually return convinced that theirs was the better showing.

Participants in the *pasola* represent both ancestral personae who confront each other across the terrain of an early ritual division and ambitious individuals who want to impress others with their horsemanship. The combat itself only gradually warms up. During the first hour or so, the best riders will not enter the field, leaving the terrain open to younger riders who let their horses run loose before the spectators. Members of one side may shout or sing taunting challenges to their adversaries: 'Where are you, the counterpart with bent knees, the partner with parted hair?', they will call out, using the couplet *'papa ndende kundo/nggaba horo longge'*, which refers to opposing sides in warfare or alliance.

The first blunt lances tossed rarely strike their mark. It is only the most experienced riders, who dare to get close to their opponents, who strike the head or flanks. Groups of five or six riders will charge towards the other side, then circle back, led by famous champions and followed by younger apprentices. Short bursts of intense activity alternate with more leisurely circling of the field, then occasional concentrated attacks. Although it is forbidden to ride straight into the opposing formation, angry riders do at times try to give chase, and hundreds of frightened horses and spectators may retreat hastily before such an assault. Falls occur most often in the middle of the battle, so the rider is in great danger of being trampelled by the other horses. Wounds from the lances have caused blindness in both horses and riders, and occasional serious injuries to the ears, throat, or skull.

Three *pasola* tournaments are held: the first one, in late afternoon at Bondo Kawango, is considered a 'training session' to 'soften the horses' feet'. The second, early the next morning near the 'trunk of Kodi', is the climax, generating a ritual heat which coincides with the rise of the sun to its highest point in the sky. Once the combat has been called abruptly to a halt by the Sea Worm priest, people return to their ancestral villages for chicken sacrifices. Then, in late afternoon, they gather again for a last battle near Tossi itself, which helps to 'cool down' the excitement of the morning and dissipate remaining aggressive feelings.

The *pasola* is interpreted by some local commentators as a substitute for war or an exercise for war, remembering the not too distant period at the beginning of this century when the island was torn by tribal warfare and mounted parties of warriors took heads from highland peoples. At the present time it expresses an aggression which is more properly interpreted as reflecting the tension within the region between hierarchical ritual order and egalitarian competitive exchange. On the *pasola* field, a certain kind of supremacy — one based on history and precedent — is challenged by another, and in many ways more pervasive, mode —

that of ostentatious achievement. Young men ride to make a name for themselves by impressing girls and prospective in-laws, and older ones display their ability to command, the following that they have among their juniors, and the fine possessions that they have acquired. All participants — both riders and spectators — come dressed in traditional ceremonial finery, with their best racehorses, decorated weapons, thick headcloths, and jewellery of gold and silver. For the first half hour of the combat, the event is, in fact, more a parade than a contest, as riders circle the field in long, easy, loping strides so that the jingling bells on their horses' halters can ring out and those watching can observe them and identify them by their colourful dress.

The display of the *pasola* is intended not only for the human audience, however, but also for a spirit audience of spectators: the deities from the sea who bring new infusions of fertility to the maturing rice crop, and encourage human reproduction through a sexually charged atmosphere of licence and ribald teasing.

Mediating between the visible and invisible audience are the Rato Nale, alternately translated as the 'Lords of the Sea Worms' or the 'Lords of the New Year'. As the highest ranking ritual authorities of the region, they control and co-ordinate all activities within the Kodi ceremonial system. The Rato Nale has the power to heal wounds or injuries, but also to punish infractions of their rules or to damage the fertility of the entire region. They 'brood' over the coming of the New Year, going into seclusion for a period of three months preceding the swarming of the worms to concentrate the sacred energies of the rice crop. This confinement is described in couplets as 'the hen brooding over her eggs, the sow calling to her young' (*a bei manu na kabukutongo taluna, a bei wyawi na karekongo anana*). In a behaviour identified as symbolically female,[3] the priest is said to be 'mourning' the death and burial of the rice goddess, and preparing for her reincarnation in the coming harvest. During this period he must refrain from eating corn or any root crops, and remains inside his house, protecting the region from lightning and heavy winds by his own immobility. In the priest's apparent weakness, however, lies a hidden strength, as his vulnerability is associated with the vulnerability of newly planted crops. The passive priest of the source villages requires the protection, and the obedience, of the more daring warriors and entrepreneurs at the periphery. The apparent equality on the playing fields is balanced by a deeper inequality which is associated with a diarchic division of power. The unmoving source of the sea worm ritual controls time and the agricultural calendar, at the same time that it is opposed to the more active, achieved powers of the centres of head-hunting, agricultural production, and competitive feasting.

The contest pits riders against each other as equals at the same time that deference is paid to the prior ritual authority of the founding villages, which control the timing of the combat and inaugurate sacrificial offerings. Thus it permits the expression of resentment against the ceremonial centre at the same time that it acknowledges the pre-eminence of a passive, unmoving authority which 'anchors' Kodi cultural identity at a single point and source.

Elements of licence and unruly behaviour are as much a part of the period of anticipation of the sea worm swarming as the restrictions and controls on the high priests. For a month before the worms arrive, no fishing is allowed on the western beaches in front of the *pasola* field, but groups of young boys and girls go down to the seashore every night to sing teasing *kawoking* songs, in which they coax the deities from the sea to come out into their traps. By the light of the full moon, they sit in groups on the sand and exchange playful verses which use the imagery of the *pasola* and the sea worms in erotically charged courtship games. The worms are told to come to 'wriggle and swarm in the trough that does not leak', along with the male counterpart, the *ipu* fish, invited to 'turn and twist in the container that does not spill'. Women are most active in composing and singing these short songs, many of which reflect tensions between the sexes through the comedy of sexual allusions. In a typical verse, the skills of a horseman on the field are evocatively juxtaposed with those of a lover:

Squeeze tightly on your Kodi stallion	Kapiriridi koyo ndara kodi
Still a little tighter there	Hodi kyapiridi kiyo
Press hard so the stallion rides straight	Kapi pandaha wadi kiyo ndara kodi
Still firmer to keep your loin-cloth from slipping	Hopi kyapi la maheria hanggingo
So the horned peaks on your headcloth will stay erect!	La maheria kadu mete!

In other verses, it is the accessories of the warrior — the tall sabre with a plume of black horsehair and the jingling bells hanging from his horse's neck — which are the object of veiled references to his virility:

Why are the bells on the horse's neck broken?	Paba ba na mbera a longgoro koko ndara?
You say the sound was piercing	Nggubu wemu

Even sharper, even stronger	Rehi liyo, rehi calo
Until they came out of the pastures	Oro loho la marada
Bogged down in the exhausting mud!	Nola njenduko haghogha!
Why did the handle of the sabre snap?	Paba ba na mbata mbekatungga pandi ceko?
You say it was once so strong	Ngguba wemu
A bearded sabre, a lordly upright sabre	Teko ndari, teko rato
Until it came from the forest of pigs' swamps!	Oro tamani kandaghu la kapore koko wawi!

The verbal taunts launched by the women play on images of horsemen who ride majestically into the battle, but return small and bent over after travelling through the pastures and forests identified with women. The taunts come along with direct possibilities for erotic liaisons to release some of these tensions. Traditionally, boys and girls were allowed an unusual degree of freedom in the last few weeks before the sea worms swarmed, often wandering off and trysting in the sands. This period was belived to be a propitious one for initiating a love affair, which would later come to fruition in the period of harvesting the mature crops of rice and corn.

The centralised complex of offerings, taboos, and calendrical observances connected to the swarming of the sea worms and the *pasola* establishes a hierarchical principle of unity which gives this ceremony precedence over all others. Within a complex and multiple division of ritual powers among villages, the *nale/pasola* festival is the only one in which no-one is excluded. The permanent ideas of precedence, order, and wholeness focused on this event contrast strongly with the reversible inequalities of wealth and status which are determined through feasting. Although both feasting and the *pasola* are competitive, one results in a cumulative series of achievements leading to a title of leadership (the horse name, the *rato* designation), while the other is always the ephemeral triumph of the champion, forced to prove himself anew at each year's contest.

As the only unifying ceremonial in which all Kodinese may participate, the *nale/pasola* festivities play a crucial role in defining Kodi as a political entity, and have historically been at the root of decisions about who would serve as political arbiters with outside powers (see Hoskins 1988a). The apparent challenge which occurs on the playing field is not so much a 'ritual of rebellion' (cf. Gluckman 1963), such as one might find in more clearly centralised states, as a playing with ideas of equality and inequality within the carefully defined parenthesis of a ritual combat. The violence and aggression which is found here is kept under control and neutralised by the overarching authority of the source villages, which are still believed to maintain a power of life and death over the participants. The *pasola* thus remains, in the final analysis, a ritual contest without winners and losers, and not a secular sport. It remains a clash of cosmic powers and dualistic oppositions rather than a simple display of skilled horses and riders, distinguishing it from increasingly popular horse-races now held in the dry season of each year.[4]

IV SACRIFICIAL GIFTS TO THE WORMS

Themes of vitality created by the intersection of male martial traditions and female values of nurturance and fertility are played out in the accompanying ritual. When people go down to the beaches to collect the sea worms and watch the *pasola*, they must also bring rice, chickens, and betel nut to be offered to the ancestors. Specific requests for male or female descendants, or announcements of changes in residence, group composition, or ceremonial procedure are also announced to the ancestral spirits along with the sacrifices of the season. Betel nut is scattered on the tombs of each family's most direct ancestors in a simple rite known as *hengapango* soon after the family's arrival, and special stores of rice set aside for this occasion may be taken down to prepare a meal of communion with one's forefathers. A chicken is sacrificed from each household in its lineage house after the first combat of the morning. The entrails of this chicken are examined by the elders of the house: if they are straight and unblemished, this augurs well for the health and prosperity of the members of the house. If they are twisted or speckled with red, there may be illness, poor harvests, or even deaths.

In the most important ritual villages, the private household sacrifices are followed by a more public meal of representatives of each cult house. The first share of chicken and meat in this meal is dedicated to the sea worms themselves, and other portions are set aside at the clan altar, the household altars, and on the tombs of important ancestors. Each participant must confess to any misdeeds which might create dissention within the village (theft, adultery, trespassing, stealing rice souls, or breaking food taboos) before he can eat the specially consecrated rice of the sea worms (*ngagha nale*).

V THE *PASOLA* IN MYTH: A NARRATIVE OF ORIGINS

The origins of *nale* and *pasola* are told in myths associated with the villages (ritual centres founded by members of a single patriclan), whose ancestors acquired original rights to them. Brought by a culture hero from overseas, the arrival of the sea worms marks the renewed fertility of each agricultural season — but their coming is associated with an apparently random series of wanderings across the territory. The stages of their journey are, however, significant in the political geography of the region, because they distribute claims to a prestigious history of contact with the centres of seasonal vitality.

The Origins of *Nale* and *Pasola*[5]

Tyemba and Ryaghe were the first Kodi men to come to Sumba. They migrated from islands to the west, sailing to the northern promontory of Sasar, where their wooden boat was smashed to pieces on the 'stone bridge' which then linked Sumba and Sumbawa. Their father was Tana Mete ('black land'), their mother was Ndabi, and their descendants were also 'black' (i.e. members of the village of Mete). They travelled to the island of Sumba with many other companions, but they were the first to reach the western tip of Kodi. They sailed around the island to the east, past the southern districts of Anakalang, Wanokaka, Lamboya, and Gaura, and eventually coming all around to reach the western beaches. They stopped in Balaghar and set up a stone, called the Tyemba Ryaghi stone. Near the western tip of the island they stopped and planted a garden. They called this place Kolo Ndako, the 'kole' board game which wanders, since they played this game (inserting *dedap* seeds into holes on a wooden platform) as they sailed.

One day they wandered further inland and discovered there were other people living in the area, who did not have gardens or cook their food. They were hunters, skilled in herbal medicine, and eaters of raw food, who knew many poisons and occult secrets. They lived with the wild animals and wild spirits of the region (*marapu la kandaghu*), and some of them were said to be witches and eaters of human flesh. Tyemba and Ryaghe made a peace pact with one of them, an old woman who lived in a hut made of bitter creepers (*warico lolo kapadu*). She had no fresh water, only brackish water, and no cured tobacco or dried areca nut, only fresh leaves and fruits. But she lived in a fertile valley, and was willing to let them live beside her if they promised not to harm her or her husband. Tyemba and Ryaghe cut off bits of their fingernails and hair and scraped a bit of flesh from their tongues to be put in a

bamboo flask as proof of their pact. They settled at the upper
end of the village, in what became Mete Deta, and the old
woman and her descendants settled in Mete Wawa. But the
village was still empty and lonely.

In Bukambero, Tyemba and Ryaghe met two others who
had migrated from the west with them in their boat — Atu
Awa and Lendu Myamba. They set off to seek other
companions, and came upon two brothers living in a monkey
shack in the forest — Mangilo was the older one, and Pokilo
the younger. The boys did not want to leave, so Tyemba and
Ryaghe tricked them by filling their drinking gourds with
coconut water and offering this to them. When they asked
where this delicious water came from, they said it came from
Kodi, and invited the boys to move there to keep them
company.

Mangilo and Pokilo first moved to Wei Walla in Bukambero,
where they were given 'sweet water' to drink every day so
they would not realise how parched the region could become
in the dry season. Lendu Myamba was reluctant to part with
the boys, so Tyemba and Ryaghe gave him a huge old
breastplate called the Marangga Bali Byapo (the couplet
name for the Kodi river basin). In return, they asked for a
'plaything' to be given to the boys to amuse them in their new
home. The toy that they most wanted to have was *nale*,
brightly coloured sea worms of all colours which would
mysteriously appear along the western beaches in February,
and should be greeted with ceremonial combat on horseback.

Lende had obtained *nale* overseas, where he had courted
the daughter of a great foreign lord (*rato ndimya, rato dawa*).
He had defeated all her other suitors at games of darts and
skill, and won the right to ask her father for the greatest gift of
all. 'Give me eternal life', he asked, 'I am tired of working so
hard to keep my gardens going in the dry, harsh climate of this
land.' 'Alas, there is no eternal life', the girl's father said, 'but
I will give you a gift of returning life, to renew the land with
fresh waters. Each year, if you receive the sea worms well,
and they are abundant, your rice harvests will be good, and
your descendants will be plentiful.' Lendu was given the sea
worms in a small trough, and he was also given a small
quantity of honey and ginger to sweeten them. He prepared
to return home, but the lord's daughter would not come as his
wife. She threw herself into the ocean instead, her body parts
breaking up into many tiny pieces with would wash up along
the beaches of Kodi in February — red pieces from her rosy,

betel-stained lips, blue and black pieces from her long flowing hair, golden pieces from her smooth skin. If he found the right spot in Kodi, he could release the sea worms from his trough and call the other worms to swarm and reconstitute the lost body of his sweetheart in the sea.

Since Lendu lived in the interior and could not release the sea worms near his home, he surrendered them to Tyembe and Ryaghe, so that they could be given to Mangilo and Pokilo, as a 'trifle or bauble' (*maghana lelu, mangguna hario*) which the young boys could play with. They took them closer to the seashore, but as the boys played, the worms got washed away in the river water. They searched for them, and found that the worms had been caught by the roots and trunk of a tree at Kawango Wulla (near the site of Tossi). The worms were rescued and taken out to the sea at Kawoto, near the tip of Kodi at Karosso. But the land there was too rocky for horses to run without hurting their feet. Then they took them to the midpoint of the coastline, Halete, where the beach was renamed Kapambalo Nale Hari, Karangga Rica Marapu ('platform for sacred sea worms, beam of spirit pule wood', the present site of the *pasola* combat). There, they swarmed in great numbers and washed up on the shore into the smaller troughs of people who came to collect them, so the sea worm woman (Inya Nale) could be reborn. And it was clear that after this long journey, the sea worms would remain at this site, which is where Pokilo and Mangilo built their own village, Tossi.

After several years, as the region grew more inhabited, settlers along the western beaches began to notice that their pigs and chickens were disappearing mysteriously. Ra Hupa, a fire-breathing figure who cast splendid multi-coloured nets with golden weights on them, was catching not only the creatures of the sea but also the creatures of the land. When others objected, he punished them with lightning or fierce winds, blowing so furiously that all their tender young crops were destroyed. Finally, the people of Kodi decided to meet in Tossi to establish a system of order which could control his rampages.

Representatives of each of the ancestral villages played each other in the traditional children's games of *kadiyo* (spinning tops) and *kalaiyo* (disc-throwing), *kule* (a board game with *dedap* seeds) and *buke* (bamboo darts). Mangilo and Pokilo defeated all the others, easily smashing their opponents' tops and outdoing them in contests of skill and strategy. This established their right to serve as the 'mother father' figures of the region.

Mangilo, the elder brother, retired to the village of Bukubani at the furthest western tip, to guard the immovable urn of sacred water, and count out the months of the year. His duty was to enforce the traditional calendar, controlling the times of planting and harvesting so that none of the tender young plants would be picked too soon. He took from Ra Hupu the 'lightning stones' which could be used to direct lightning bolts at offenders of these prohibitions, and promised to remain in strict confinement in the months preceding the coming of the sea worms — concentrating the fertile energies of the region 'like a hen brooding over her eggs, like a sow calling to her young' (*na kabukutongo taluna, na karekongo anana*). He thus became the first Sea Worm Priest (Rato Nale).

Pokilo, his younger brother, stayed in Tossi as the 'horse with an erect tail, the dog with a black tongue' (*ndara ndende kiku, bangga mete lama*), the executive power and 'traditional policeman' who rode from one village to another to make sure the land boundaries and calendrical rules were kept. His descendants were also Sea Worm Priests.

Ra Hupu was banished to the other side of the river, to 'soak his head and cool his liver' in the fresh waters which flowed down from the highlands. He was forbidden to steal from his fellow Kodinese, but his fierceness and fire were given a new focus in the skull tree erected in the centre of his village, where the heads of enemy highlanders could be hung. On the warpath, his magical weapons and control of the winds could once again be used — but in Kodi they would not be unleashed for as long as the Sea Worm Priest (Rato Nale) remained confined during the month awaiting the swarming of the worms. Ra Hupu became the adopted nephew of Pola Kawata, a python deity who came originally from the highlands but visited Kodi with abundant rains and unusual fertility in certain years.

VI ANALYTIC COMMENTARY

This narrative establishes a series of processes which repeat themselves in the story:

(*1*) The sea worms are an 'entertainment' which must be 'won' through competitive games — first by Lendu on his voyage overseas, then by Mangilo and Pokilo. (In a later narrative, the same games were also played to transfer the rights to *nale* to the more distant valley of Balaghar.) The playing of other, smaller games (the board game and discus game) along the way reinforces the general idea that games of chance and skill are, in fact, forms of divination, as the results are guided by the invisible hands of ancestral authorities. The classification of these fertility-bearing creatures as children's toys emphasises both the light-hearted, ludic element which pervades the atmosphere at the *pasola* and the youth of these early forebears, presenting a version of the development of Kodi society modelled on the life-cycle.

(2) Persons and objects are transferred along with gifts which provide both a testimony to their origins and an enduring obligation to reciprocate in the future. The gift of the *nale* from Lendu is marked by the countergift of a large gold breastplate, which can still be seen in Bukambero, presented by the local people as 'proof' that they had earlier rights to the sea worms. The rights to the worms also conveyed rights to the land, which was represented by the large breastplate which gave its name to the present territory.

(3) The passive ritual power of the elder, symbolically female brother is separated from the more active, warlike powers of the younger brother. The exile of Ra Hupu more or less repeats this pattern, as the potentially disruptive, violent rampages of a rebel with magical powers are brought under control through a pact kept with the ritual centre. Once again, passivity is the price paid for deference, and by surrendering his own mobility the Priest of the Sea Worms gains a greater authority. Ra Hupu is re-integrated into the group, but kept at a distance, his destructive urges redirected towards enemies, but still the source of new powers.

(4) The recitation of place-names throughout this narrative creates a chain of related spaces which are ordered through notions of ritual precedence. At each point where the sea worms were released, the people of that area acquired residual rights to play a role in sea worm ceremonial, and left physical evidence of their passing (in the form of rocks, heirlooms, changes in the geographic formation). Thus, the coming of the sea worms is spatially anchored in a series of inland villages, whose resources may then be drawn on in mounting the coastal festivities.

(5) The origin of *nale* from the body of the lord's daughter recalls a more widespread mythic tradition about the origin of rice and other garden crops from the body of the rice goddess. In Kodi, Mbiri Koni is the name of the ancestress who was sacrificed by her father, buried, and then 'resurrected' in the new crop of rice which sprouted four days after her death. In some versions, she was just a baby, in others she is nubile young woman or even already

married. Sometimes, specific body parts are detailed as the source
of particular garden crops (i.e. corn from her teeth, bananas from
her heart, coconuts from her head). Both stories share themes of
female sacrifice and resurrection, and associate the gift of fertility
with the death of a daughter. Mbiri Koni is sometimes identified as
the daughter of Pala Kawata, the python in the highlands who
adopted Ra Hupu as his nephew.

(6) The *pasola* itself is assumed as a necessary accompaniment
to the sea worms, and no mythical justification is given for the use
of horses or the rules of the game. Informants generally maintain
that horses have always been a part of local culture, and the first
ancestors arrived riding them.[6]

VII THE *PASOLA* IN THE CEREMONIAL SYSTEM

The myth of the origin of *nale* and the *pasola* serves as a charter for
the ceremonial system. The moral relationship that binds siblings
together on unequal terms establishes a tension between the
polarities of passive containment and active involvement. The
division and distribution of ritual tasks repeats this tension in a
series of four other paired oppositions: (1) *kahale/katoda*, agricultural and head-hunting rites, 'the harvest of life and death'; (2)
nale/padu, the period of permissiveness and licence when the sea
worms swarm versus the period of silence and prohibitions; (3)
kabalyako/opongo, the powers of the sky and earth, lightning and
earthquakes; (4) *ngguhi/ndara halato*, the immobile urn of
governmental authority and the roaming horse of boundary
enforcement.

In each case, the first term of these pairs is associated with life,
fertility and positive forces, while the second is linked to death,
scarcity, and negative sanctions. In the sacred geography of Kodi,
the first term is always associated with the centre, while the second
is found at the periphery. The positive manifests plenitude by
inertia. Since its potency is assured, there is no use of force or
constraint. People are irresistibly attracted to the centre and thus it
sustains itself without effort. At the periphery, an intensity of
achievement and activity is required. Valeri's (1990) discussion of
the relation of history and diarchy in Polynesia illustrates this
contrast well. In a pattern which is widespread in the Pacific, the
paradigmatic younger brother excels 'in activities that imply
hierarchical instability and reversibility — games, sports, and
ultimately war. He always beats his elders at these, thereby
demonstrating the anti-structural and anti-hierarchical connotations of his vital power. These are further demonstrated by the
fact that he is often a 'trickster' who is able to violate with

impunity the rules respected by his elders and on the respect of which their authority is based' (Valeri 1990: 10).

Lendu, the culture hero who first obtained *nale*, Rato Pokilo, who uses tree sap to make his dart stick to the target, and Ra Hupu, who steals and wreaks havoc with the possessions of his fellows, are all mythological figures of juniority and the periphery. In contrast to them, Tyemba and Rato Mangilo are passive authority figures, who show the extent of their power by delegating its execution to younger subordinates. The long journey of the sea worms follows them as they change hands from young adventurers to community elders, who finally entrust them to the older of the two brothers, drawn as if by magnetic attraction to the roots of a large tree planted at the centre of the line of ancestral villages.

The unequal relationship of siblingship provides the paradigm for other ceremonial divisions, in which a senior, immobile term cedes some authority to a junior, active younger brother. Within the original 'trunk of Kodi', Tossi remains the centre for agricultural ritual (*padu mono kaba*, 'bitter and bland crops') and governmental authority. Bukubani controls the lightning stones, and Bondo Kodi has the skull tree of head-hunting. Mete, the village descended from the first indigenous inhabitants, has domain over earthquakes.

After the first generation of heroes worked out this division of powers, a later series of historical narratives deals with the demographic expansion of the population and the extension of these divisions to the regions of Bangedo and Balaghar. The concentration of the most sacred powers at the centre is continued in spite of this expansion by making each delegation of power also, to a certain extent, a diminution and dilution of the original. Thus, the rights to head-hunting and warfare ritual were transferred by transplanting a sapling of the skull tree in Bondokodi to four villages in Mbali Hangali ('the other side of the embankment'): Toda, Ndelo, Kere Tana, and Bongu. A few pebbles from the lighning stones of Bukubani were moved to Wainjoko in Balaghar, but its power to control rainfall was seen as derivative from the original source. The elder village retained the prestige of priority and precedence, and its successors in other valleys were told to remember that they are only the 'younger siblings and children' (*ari ana*) of the founding power.

The hierarchical chain of transmission is historical rather than mythical: elders in the villages concerned can provide specific genealogies for the ownership and acquisition of the heirloon objects associated with the transfer of powers. Thus, while the

Kodinese see as invariable the basic pattern of relationships between elder brothers and younger brothers, and the sense of mutual obligation implied by that balance of power, they do not see their ancestral villages as part of a timeless order. The present distribution of ritual powers is the result of a long historical development, recorded in complex successions of names revealing a process of segmentation and dispersal accompanied by territorial expansion.

The *pasola* has also spread from Kodi to three other districts in West Sumba. It is currently performed in Lamboya on the morning that the sea worms swarm in February (usually the same day as in Kodi), and Wanokaka and Gaura in March, for the second swarming of the worms. Mythic traditions in Wanokaka trace the transfer of the ritual to negotiations concerning a local woman who eloped with a Kodi man, and refused to return to her homeland. Her grieving husband went on a long journey to search for her. To keep her as his wife, the new husband in Kodi paid a full brideprice in gold and livestock, adding the sea worms as a final 'gift of life' which would replace the reproductive powers of the woman he had taken away (Mitchell 1981; Shaw 1976). The pattern of the Kodi myth is inverted: When the sea worms were first brought from overseas, they were supposed to accompany the bride, but ended up a substitute for her broken body. In the transfer to Wanokaka, the worms themselves travelled to another district as a 'female valuable' which took the place of the reluctant bride. In both cases, however, the worms were an alternative form of fertility and vitality, associated with a beautiful but unattainable woman and transferred in her absence as a kind of compensation.[7]

VIII THE *PASOLA* IN HISTORY: WARFARE, PACIFICATION, AND TOURISM

Kodinese accounts of the *pasola*'s origins and transfer are presented in a historical as well as a mythical context, as we have seen, since they recognise diversity and foreign influences. Comparisons with ritual combats in other parts of Indonesia may lead us to sharpen the focus on how such performances are linked to cosmic oppositions, earlier patterns of feuding and warfare, and present efforts to stage such events as part of a packaged 'folklore' of local culture, suitable for tourists and government dignitaries.

The mythic assertion that *pasola* was introduced from a splendid kingdom to the west may have some justification. Jousting on horseback is mentioned in the Malay epic *Hikyat Hang Tuah* as part of the entertainment at the Javanese court of Majapahit. The first detailed description of the weekly tournaments (*senenan*) in

the squares near the royal citadel comes from a Dutch account of
the events in Tuban and at many other courts in central and
eastern Java. Reid's (1988: 187) portrait of these contests
highlights similarities to the present-day *pasola*:

> About four in the afternoon the younger braves of the court
> would converge on the square after parading through the city
> on their magnificently attired horses. There they would
> engage in a series of charges and manœuvres, one generally
> pursuing the other down the length of the field, with the aim
> of knocking each other off their horses with blunted spears. In
> reality, this happened seldom, and most attention was paid to
> the horsemanship displayed in the constant wheeling and
> turning on the square. The king was always present for these
> occasions and, at least in Mataram, took part in the jousting.

Reid interprets these tournaments as a 'metaphor of war in which
young aristocrats proved and displayed their qualities' (Reid 1988:
187). Before 1600 these tournaments seem to have been quite
bloody, as provisions are mentioned for the families of victims (Ma
Huan 1433: 94, cited in Reid 1988: 187). Tome Pires described the
'knights' of Java (*cavaleiros*) as proud horseman who often
provoked deadly combat on an individual basis; 'The noblemen
are much in the habit of challenging each other to duels, and they
kill each other over their quarrels, and this is the custom of the
country. Some of them kill themselves on horseback, and some on
foot, according to what they have decided' (1515: 174–96, cf. Reid
1988: 187).

Many Sumbanese noble families like to imagine that they are
descended from aristocratic refugees from the royal courts of
Majapahit, and have thus embraced interpretations which trace
the origin of the *pasola* to Java. Some educated Kodinese, aware
of the importance of the Javanese cult of the goddess of the south
seas, Nyale Loro Kidul, have suggested that this is the source of
the term for *nale*, but there is no evidence that jousting tournaments
were linked to sea rituals in Java. On Lombok, courtship songs are
sung in a festival atmosphere to greet the swarming of the same
worms, so there has also been speculation that the *nale/pasola*
complex is linked to these celebrations (Ecklund 1979).

On Savu, a small island just southeast of Sumba, the swarming
of the sea worms is ritually observed, and used to name months
within the lunar calendar (Fox 1979). As on Sumba, each domain
has a separate calendar of named lunar months that coincides
with the seasonal agricultural activity. The biannual swarming of
the sea worms is used to check a system of priestly intercalations
which bring the named months roughly into line with the transition

between the monsoons, and the growing cycle of rice and corn (on Sumba) or the green gram and lontar palm (on Savu). On both islands, the end of one year and the transition to a new one is accompanied by a spectacular ritual combat.

In West Sumba, the *pasola* in February (for Kodi and Lamboya) and March (for Wanokaka and Gaura) is timed to coincide with the swarming of the sea worms, and seen as an 'entertainment' for them. On Savu, large-scale rock-throwing battles are held in September–November, before the season of the sea worms in December–February, but similar beliefs accompany the combat: 'It is a month when the spirits are said to rise from the sea and must, in the end, be driven from the villages. The final pig sacrifices for anyone who has died during the year are held in this month prior to the renewal of the earth' (Fox 1979: 155). Ritual rock-throwing combats in Savu close off the high ceremonial season, which is inaugurated just after the green gram harvest in celebrations which climax with a sacred cock-fight. Since *nale* and *pasola* also signal a 're-opening' of the ceremonial season after a ritual silence of many months, the structural position of the cock-fight within the Savunese calendar is most similar to that of Sumba's mounted combat.

The coupling of human and animal combat in many different locations suggests that the blood of one has often served to represent the blood of the other. Thus, in Java jousting tournaments declined as warfare became less frequent, and were gradually replaced with animal combats (fighting elephants, tigers, buffalo, and oxen) (Reid 1988: 1987). The Savunese give two explanations for their cock-fighting, which bring it close to the *pasola*: 'It is a struggle between classes of spirits represented by men, the spirits of the land and the spirits of the sea; and it is a substitute for warfare between clans and villages' (Fox 1979: 165).

Parallels with ritual combats in other parts of Indonesia may also suggest how the ritual significance of the performance is connected to its recent history, and involvement in local rivalries.

The finest riders at the *pasola* still tend to come from Tossi, the 'mother father village' of the whole region, and now from those more recently founded villages on the other side of the river which have emerged as Tossi's rivals in a wider political context. In the 1970s and 1980s the informal leader of Tossi's horseman was Tari Nggoko, a descendant of the first Kodi raja. Rangga Baki and Ratenggaro, the home of the third Kodi raja and a famous centre of head-hunting rites, are now the main challengers from Bangedo.

The fathers and grandfathers of contemporary riders were horsemen whose skills were valued not only as entertainment (for both humans and spirits) but also as part of the practical arsenal of warfare. In the late nineteenth century Rangga Baki and Ratenggaro were locked in a deadly pattern of murders, feuding, and livestock raiding. The two villages, however, suspended their hostilities at *pasola* time, when they rode together against their opponents from Tossi and source villages of the 'trunk of Kodi'. One of the region's main *pasola* champions and fiercest warriors, Rato Muda of Rangga Baki, was said to have been approached to serve in the colonial administration. He refused, saying 'I still have my brother's blood to avenge' when emissaries asked him to agree to serve as a peace-keeper under Dutch rule.

The usual season for headhunting, local feuding, and warfare was in July to September, after the harvest and towards the end of the dry season, when agricultural activity was at a minimum. *Pasola* thus falls at the other polarity of the calendar, in a time of agricultural co-operation, when people live in dispersed settlements near their interior gardens so they can weed and harvest the crops of the rainy season. It could perhaps be surmised that the 'peaceful conflict' of this ritual contest came about because people flocked to the beaches to collect sea worms at this time, but were unprepared for warfare.

'Pacification' was attempted in 1909, by gathering Kodi elders together to choose a native representative as district administrator, but in 1911 the representative chosen, Rato Loghe Kanduyo of Tossi, was implicated in an armed rebellion against the Dutch forces led by members of one of the headhunting villages (see Hoskins 1987c). Guerrilla warfare between Kodi warriors and Dutch forces continued for two years before the rebels were finally convinced to surrender and were sent into exile. Perhaps hoping that it would become a true 'substitute for war', the Dutch colonial administration then chose to 'develop' the *pasola* contest, and encouraged a broader participation from all over the Kodi district. When regional raiding was effectively suppressed, spectators and riders could come from much further away, and the contest began to flourish as a 'ceremony' just as some of its practical usefulness as a training ground for mounted warfare began to diminish.

Sumba is similar to the societies of early Greece and Sparta, where athletic champions were often political and military leaders as well. This seems to have been changed somewhat with the suppression of inter-regional warfare in the 1920s, and present government policies to control feasting. In effect, an informal and competitive political system is being increasingly replaced by an externally appointed administrative hierarchy where skill in horsemanship is no longer an important factor. Thus, the *pasola* is being promoted now as a folkloric celebration and a way of entertaining visiting dignitaries, rather than a religious rite and the basis of the Kodi polity.

The *pasola* could be said to be similar to many other mock-combats found throughout Indonesia, surveyed in Downs's essay (1955) which argues that these forms, as well as headhunting, had their origins in an early form of dual organisation. 'Often they would seem to have degenerated into mere sporting events indulged in at the time of general festivities, without any consciousness on the part of the participants of their religious significance' (Downs 1955: 55). While it is not clear that the *pasola* is really 'a ritual struggle representing the two halves of the universe' (Downs 1955: 59), it was clearly, and continues to be, a struggle of opposing principles. The layering of meanings — the rivalrous combat still couched within a recognition of hierarchy, the mythological precedents still interwoven with present tensions — assure that it cannot be so quickly reduced to the equivalent of present-day secular horse-races.

IX CONCLUSIONS: DUALISM AND CONFLICT IN THE *PASOLA*

Through both myth and history, the performance of the *pasola* has combined elements of competition and equality with others of unity and hierarchy. Interpretations of the basis of ritualised combat which stress a dualistic foundation play on the ambiguities of dualism itself. In his classic treatment of the process of social differentiation and disintegration in *Naven* (1958), Gregory Bateson argued that dualism could be divided into two types: a 'direct dualism', where everything has a sibling; and a 'diagonal dualism', where everything has a symmetrical counterpart (Bateson 1958: 235). One type is based on a duality within identities; the second on dualities within difference.

'Direct dualism' is the basis of complementary dyads, a grouping of relationships into artificial pairs which together suggest an idea of completion. Symmetrical dualism, on the other hand, presumes both equality and opposition. The *pasola* and the sea worm festivals which accompany it contain elements of both modes of thinking. A complementarity is expressed between the male and female roles, the exhibitionism of the riders on the field, and the ways in which they are taunted into action by ribald female songs and spectatorship. This kind of dyad is also found in the complementarity of the priests and the horsemen, as sacred representatives of the ritual centre offer a passive mode of leadership and control, which contrasts with the active displays of the riders. On the playing field, each participant is in a symmetrical relationship to the others. They confront each other on an equal footing, as symbolic equivalents who realise a distinctive individual identity through skill, bravery, and theatrical flare — but not through a different structural position.

In the world of Sumbanese mythology and political realities, brothers are always ranked, and younger/elder division introduces an 'asymmetrical dimension' which skews the distribution and projects these apparently simple relations onto an axis of complementary opposition. The quickness, agility, and daring of the *pasola* rider is presented as characteristic of the younger brother, while the stability, dependability, and authority of the elder brother is elaborated in contrast. As among the Iatmul, the pairings of direct dualism are necessarily unequal, showing elements of the process Bateson has analysed in the following terms:

> Throughout the whole field in which direct dualist thought is recognisable, it is accompanied by the thought that one of the units is senior to the other. There is no such concept in the case of persons identified by diagonal dualism and it would seem that such persons are nominally equal in status and always of the same sex; while those who are directly identified can never be equal but must differ either in seniority or sex.
> [1958: 243]

The reduplicative likeness of equivalent dyads differs from the complementarity of direct dualism in that the two are immediately placed in the same class, instead of being linked by a more obscure hierarchical principle.[7] Bateson had described this as the difference between symmetrical and complementary schismogenesis: in the symmetrical model, both sides exhibit competitively similar behaviour (i.e. riding across the field, taunting their opponents, boasting, throwing lances) and progressive change will occur as each actor reacts identically but increases the degree of reaction. In the complementary model, one side in an interactive pair takes on a role which differs but coincides with the other — being submissive in the face of dominance, offering exhibitionism to passive spectatorship. Men who rise in the *pasola* are stimulated by female attentions to display their speed and vitality, but all this dominance behaviour is also put at the service of a rite of female fertility (expressed in the sea worms and the rice taboos), and in some ways subservient to the more encompassing authority of the passive, symbolically female priests.

The coexistence of these two types of dualism within the *pasola* signals an aspect of ritual contests which is related to what anthropologists have often described as their 'integrative'

functions: an excess of symmetrical rivalry may trigger complementary rituals, this bringing a society back to at least a temporary
semblance of a steady state (cf. Bateson 1958: 290). In contests
such as this one, the frame of a calendrical repetition is used to
neutralise the potential violence of earlier oppositions, turning an
aggressive and warlike mode of behaviour into a display of unity.

Tensions between the sexes surface in the *kawoking* songs,
where women mock male activities and male pretensions, pointing
out how quickly they are diminished after contact with women.
These taunts are particularly barbed because they are sung at a
time of ritual licence, when young boys and girls are allowed an
unusual degree of freedom and expected to act in accordance with
the regional celebration of productive and reproductive forces.
The *nale* festival presents a conjoining of male and female where
the complex centred on female symbols of rice, sea worms, and
fertility is brought to meet male symbols of horsemanship, skills in
warfare, and virility. The songs which anticipate the *pasola* show
an insolent lack of respect for male spheres of achievement, and
serve to remind the men that all this prancing about on stallions
must be followed by sacrificial offerings to the deity addressed as
Inya Nale ('Mother Sea Worm'). The complementarity between
men and women is thus contested and challenged, but only to a
certain extent. I would agree with Turner's analysis of 'rituals of
status reversal' that such apparent reversals are transitory, and
finally reaffirm those categories 'that are considered to be
axiomatic and unchanging both in essence and in relationships to
one another' (1969: 176). His argument explains how such
ritualised hostility can actually serve to reinforce a pattern of
complementary but fundamentally unequal cooperation:

> Cognitively, nothing underlines regularity so well as absurdity
> or paradox. Emotionally, nothing satisfies as much as extrava-
> gant or temporarily permitted illicit behavior. Rituals of
> status reversal accommodate both aspects. [1969: 176]

Elements of status reversals in gender and political status must
also be understood within their temporal sequence, since this is
where the fragile hierarchy is established.

The *nale* season begins with months of prohibitions, restrictions,
and ritual silence, where joyous, noisy, behaviour is constrained in
order to concentrate the population's attention on the activities of
planting. The priests, descendants of the elder brother, bear the
heaviest taboos, and must remain immobilised within their sacred
houses, while the descendants of the younger brother may roam a
bit more, training their horses for the combat but not allowing
them to run across the sacred field of combat or the sands near the

sea. Breaking any of the first set of taboos on the priests would bring on dangers to the coming rice harvest (heavy winds, lightning, thunder which would 'shake the tender young grains off the stalk'). Breaking any of the second set of taboos would be personally damaging to the individual violators, but not to the region as a whole. Thus, the hierarchically superior elder descent line is identified with the unified territory (as one would expect, cf. Dumont 1977), while the inferior junior descent line represents the more particular concerns of its separate members.

As the time of the combat approaches, the complementarity of senior and junior houses is contrasted with male/female complementarity, with women using the idiom of swarming of the sea worms for an irrelevant commentary on the male prestige system. The combat itself glorifies male qualities of bravery, flamboyance, and daring — but only within the context of a ritualised confrontation. Then, once the climax of ritual 'heat' is reached at the combat of noon on the second day, the playing is stopped by the Sea Worm Priest and all participants are told to return to their ancestral villages to sacrifice chickens to their ancestors. These sacrifices are accompanied by requests for 'cool waters, refreshing waters' (*wei myaringi, wei malala*, generally translated as 'blessings', Indonesian *berkat*) from the village founders, taking the form of abundant harvests and many new human descendants. Male heat is 'cooled' by symbolically female ritual officers, and the excitement of the play is transformed into an injection of fertility and vitality for the whole region.

To the extent that the *pasola* really provides an 'alternative' to warfare or feuding, it might be said to defuse the mystique of activities by their presentation as 'sport'. I am myself more inclined to the view that the Sumbanese see the event as a training session (Kodi *doru witti ndara*, 'a softening and nimbling of the horses' feet', often translated with the Indonesian *latihan*, 'practice, exercise') for such activities, in case the need were to arise again. This interpretation is supported by the fact that the *pasola* frequently provides the occasion for outbursts of real violence between parties which are at odds with each other, and thus is now heavily patrolled by the local police and army. In 1980 a rock-fight broke out between people from the villages of Rangga Baki and Tossi. In 1981 swords were bared and one person was struck with the edge of a bush knife in a fight that occurred at the edges of the field. In 1988 a murder occurred just off the playing field in Wanokaka, and I have heard many other stories of such disturbances in addition to the above events that I have witnessed. Such individual violence is, of course, prohibited by the rules of the game but — like the brawls that erupt between players or spectators at football games — is expected and hardly a structural anomaly.

'Ritualised' violence is still, in a very real sense, actual violence, and the ceremonial frame which should serve to contain aggressive behaviour also promotes and encourages it. Male exhibitionism is intensified by the undercurrent of mockery which we have found in female spectatorship. The daring of representatives of the peripheral, 'younger brother' villages is exacerbated by the constraints of timing and ritual control which are all invested in the 'mother/ father' village of Tossi-Bukubani. So the *pasola* represents, in my view, a chafing under certain types of authority, and an opposition of principles of achievement and ascription, but not a real challenge of either.

The analysis begun to explain the tensions between human participants can also be extended to explain how the same set of relations encompasses a triad of different entities — men, horses, and gods. The *pasola* uses the form of a ritualised combat — a 'metaphor' for warfare which substitutes for some (but not all) of the functions of earlier feuding and headhunting. Horses run across the field as metaphoric extensions of their master's identity, and it is the fine-tuning of interaction between man and horse — the coordination of speed and intention, skill and co-operation — which assures an impressive performance.

The training of young animals for the *pasola* field parallels the functions of 'training' young warriors for the battlefield. The subjugation of domestic animals is a form of inequality where an inferior but physically more powerful being is brought to obey the will of a superior but smaller one, because of the social skills men have learned to control the beasts. At the same time, Sumbanese assert that man/horse relations are ruled by an apparently contradictory theme of equality — the man and his mount 'share the same fate' and travel together as companions along the pathway of life.

In a somewhat similar fashion, the division of ritual tasks between elder and younger brothers plays out a dream of superior ritual and social authority which tames and domesticates a wild and powerful physical energy, bringing it into the service of communal goals of fertility. On the playing field itself, this inequality is balanced by the frame of an athletic contest, where horses and riders from all villages can compete as equals.

One could almost assert that Sumbanese vacillate between a view of horses and men as almost siblings (and thus part of a system of symmetrical dualism) and as ranked, and thus potentially complementary, beings. In the last analysis, however, horses remain the inferiors of men, just as men remain the inferiors of the *marapu*, spirits and deities who are described as 'herding them into the corral at night, taking them out to the fields by day'. The pastoral metaphor draws together men, horses, and gods in a series of rituals where their identities are fused and then separated, matched in combat and set apart in the more enduring cosmic order.

Dualities are thus nuanced in their expression: not merely agonistic but also complementary, based on a clear hostility which also reacts to interdependence. The *pasola* moves from equal to unequal contests, from a political system of clearly competitive struggles to one where wider considerations can absorb some of their vigour. In fighting fiercely against each other, the Kodinese acknowledge a mutual origin and a set of shared values. Only through this violent and dangerous game can a ritual equilibrium be restored, and an idea of an unstable but essential cosmic balance maintained.

ACKNOWLEDGEMENT

Doctoral research in 1979–81 was supported by the Fulbright Commission, the Social Science Research Council, and the National Science Foundation, under the auspices of the Indonesian Academy of Sciences (LIPI) and Biro Penelitian, Universitas Nusa Cendana. Return visits in 1984, 1985, and 1986 were funded by the Research School of Pacific Studies, Australian National University, and the Faculty Research and Innovation Fund of the University of Southern California. Six months of additional fieldwork on ritual communication in 1988 were funded by NSF Grant No. BMS 8704498. All of this help is gratefully acknowledged, and I would like to give additional thanks to Ra Holo, the Rato Nale of Tossi, and Valerio Valeri for helping me to understand this event.

NOTES

1 The complexity of the calculation of the date of the worms' arrival arises from the fact that the traditional Kodi lunar calendar of named months must be intercalated with the solar calendar of seasonal activities such as planting and harvesting crops of rice and corn. The Rato Nale, or Sea Worm Priest, must shift the 'counted out' date of the *pasola* every four years, to allow for the slippage between the two calendrical systems which the Gregorian calendar deals with through Leap Year. Interviews with the priests in 1988 indicated the only astronimical knowledge used is the rising of the Pleiades at the beginning of the rainy season. In addition, a social calculation is made to start the 'bitter

months' (*padu*) of ritual silence so they will coincide with the period of planting before the onset of rains. The priest is the custodian of the traditional calendar, but his ability to 'control time' and co-ordinate harvest and planting activities is put to the test by his success in predicting the arrival of the sea worms. A priest who repeatedly misses the date may be ousted from his post, if a divination reveals that he is not correctly following the wishes of the ancestors.

2 The *ndara nale*, 'horse of the sea worms', is a fine stallion selected from the herd that has belonged to Tossi, the centre of the sea worm cult. It cannot be bought or traded from the outside, and must observe the same food taboos as its master. During the period of awaiting the worms' arrival, it is confined to the village, tied under the lineage cult house dedicated to the sea worms, and cannot eat corn, tubers, or any other grain besides rice. The Sea Worm Horse does not participate in the combat itself, but starts and finishes each jousting session by circling the *pasola* field. Once the combat is finished, the horse must be ritually 'cooled down' with a blessing of sacred water. The horse is believed to have absorbed the violence of the combat in the course of offering protection to other riders involved in the combat. The healing waters stored in a special ancestral urn are sprinkled on the horse's head and withers with a *kambukelo* leaf, the same leaf used to apply medicines to wounded horses and riders, and to feed the last meal of rice to a departing ghost at the end of the funeral rites. Its use indicates a separation, a movement away from intense, heightened 'ritual time' back into the secular rhythm of daily activities.

3 The three Sea Worm Priests in the ritual centres of Tossi, Bukubani, and Weingyali that I encountered in fieldwork in the 1970s and 1980s were male, but the post has also been held by a woman. Mbiri Kyoni, the head priestess of Tossi at the time of Dutch contact in 1911, and through the period of the Japanese occupation, was literally as well as symbolically female. She succeeded her husband in the position, and was famous for the accuracy with which she predicted the arrival of the sea worms, thus correctly adjusting the lunar calendar to the solar cycle of seasons. Her death in the 1940s was followed by a long drought and famine, further reinforcing the popular apprehension of her power as intimately connected to the growth cycle of rice and corn. Many informants insisted that present-day occupants of the priestly office were not as knowledgeable as their famous female predecessor. Priests are selected by divination from the occupants of the cult house of the sea worms, but many seemed to think it unlikely that another woman would be chosen. It is possible that the gender

values associated with the position have shifted in response
to historical forces and Western ideas of the separation of
church and state (see Hoskins 1988a).

4 Horse-races are now held, legally, in the regency capital of
Waikabubak in August and, illegally, in association with
gambling throughout much of the year. The appeal of
horse-racing on Sumba is quite similar to the appeal of
cockfighting in Bali, as presented in Geertz's famous essay
(1973): it is a competitive arena for evaluating status and
risking reputation, sharply distinguished from the more
controlled and ceremonial contests of ritual.

5 This narrative is a composite of accounts taken from
informants in the villages of Mete, Wei Walla (Bukambero), Bondo Kodi, and Tossi, since each retained the
rights to tell his portion, even when all agreed on the basic
structure of the tale. It provides a fuller version of the myth
recorded by Onvlee (n.d.) in 1932 in interviews with
Haghi Tena and Rehi Kyaka Ndari of Mete and Tossi
respectively. Portions of the text collected in Mete have
also been published in Hoskins (1988b).

6 Historically, it seems that domesticated horses had spread
as far as Sumba by the early fifteenth century, perhaps
introduced by Arab and Portuguese traders during the
period of sandalwood exports through Bima (Reid 1983:
34). However, little is known definitively about how long
the animals have been on the island.

7 After reading a version of this paper, Edgar Keller
provided some interesting comparative materials from
Lamboya. Although the pasola is publicly linked to the
appearance of seaworms in February, in fact it is most often
the case that there are no seaworms along the beaches at
that time to be collected. The big swarming does usually
occur in March, but because of a mistake made by priests at
the ritual center of Sodan, 'the ancestral spirits sentenced
the Lamboyans to perform the rituals of the *nale* month
without the presence of the seaworms' (pers. comm.).
Instead, elders maintain that the pasola is held as an
entertainment for the spirits who will send the rains. The
pasola is connected to a combat on foot called the *pahalana*
which occurs in Lamboya in connection with the 'bitter
month' festivities (*paddu*) in November. Thus it would
seem that these ritual combats do not always coincide with
the appearance of the sea worms. As in Wanukaka and
Gaura, the people of Lamboya trace the origin of the
pasola and nale festivities to Kodi. I might also add that in
Kodi although the arrival of the sea worms is *predicted* by
the priests to occur in the month we call February, at times
the larger swarming occurs in March. This is called the *nale
wallu* or 'leftover sea worms' in Kodi, and during the

period 1979–1988 the majority of the worms appeared one month later in 1980, 1984 and 1988. It is accepted that the priests' predictions are generally correct if the worms make their appearance seven days after the full moon in either of the two *nale* months, *nale bokolo* ('greater sea worms') or *nale wallu*.

8 Needham's reconsideration of these notions of opposition argues that Bateson's predilection for spatial conceptions of dualism trapped him in a geometric analogy which 'did not well serve the purpose of translation for which it had been adopted' (1987: 191). Since my concern here is with the *different moral relationships* which bind dualistic pairs rather than with the issue of their abstract representation, I find the geometric analogy in fact rather apt. From this perspective, the 'directness' of one type of dualism corresponds to the propensity of members of such dyads to be paired in ritual couplets, where their differences are directly compared. The 'diagonal' relationship of riders on the *pasola* field, who oppose each other from symmetrical battle lines and ride to meet in the centre, allows for a confrontation which produces momentary victors. This contrasts with the complementarity of male and female or senior and junior, where confrontation is not appropriate, and inequality is masked by interdependence.

REFERENCES

Bateson, Gregory (1958 [2nd edition]). *Naven: A Survey of the Problems Suggested by a Composite Picture of the Culture of a New Guinea Tribe Drawn from Three Points of View*. Stanford: Stanford University Press.

Downs, R. E. (1955). Headhunting in Indonesia. *Bijdragen tot de Taal-, Land- en Volkenkunde* 111, 40–70.

Dumont, Louis (1977). *Homo aequalis: I. Genèse et épanouissement de l'idéologie economique*. Paris: Gallimard.

Ecklund, Judith (1979). Marriage, Sea Worms and Song: Ritualized Responses to Social Changes on Lombok, Indonesia. Unpublished Ph.D. dissertation, Cornell University, Ithaca.

Fox, James J. (1979). The Ceremonial System of Savu. In *The Imagination of Reality*, eds A.A. Becker and A.L. Yengoyan, pp. 145–73. Norwood, N.J.: Ablex.

—— (1987). Southeast Asia: Insular Traditions. In *The Encyclopedia of Religion*, ed. M. Eliade, pp. 520–7. New York and London: Macmillan and Collier Macmillan.

Geertz, Clifford (1973). Deep Play: Notes on the Balinese Cockfight. In *idem*, *The Interpretation of Cultures*, pp. 459–90. New York: Basic Books.

Gluckman, Max (1963). *Order and Rebellion in Tribal Africa*. London: Cohen and West.

Hoskins, Janet (1986). So My Name Shall Live: Stone-Dragging and Grave-Building in Kodi, West Sumba. *Bijdragen tot de Taal-, Land- en Volkenkunde* 142, 31–51.

—— (1987a). Entering the Bitter House: Spirit Worship and Conversion in West Sumba. In *Indonesian Religions in Transition*, eds R. Smith Kipp and S. Rogers, pp. 136–60. Tucson: University of Arizona Press.

—— (1987b). Complementarity in this World and the Next: Gender and Agency in Kodi Mortuary Ceremonies. In *Dealing with Inequality: Analysing Gender in Melanesia and Beyond*, ed. M. Strathern, pp. 174–206. Cambridge: Cambridge University Press.

—— (1987c). The Headhunter as Hero: Local Traditions Reinterpreted as National History. *American Ethnologist* 14/4, 605–22.

—— (1988a). Matriarchy and Diarchy: Indonesian Variations on the Domestication of the Savage Woman. In *Myths of Matriarchy Reconsidered*, ed. D. Gewertz, pp. 34–57. Sydney: University of Sydney Press.

—— (1988b). The Lips Told to Speak, the Mouths Told to Pronounce: Etiquette in Kodi Spirit Communication. In *To Speak In Pairs: Essays on the Ritual Languages of Eastern Indonesia*, ed. James J. Fox, pp. 29–63. Cambridge: Cambridge University Press.

—— (1989). Losing and Getting a Head: Warfare, Exchange and Alliance on a Changing Sumba 1888–1988. *American Ethnologist* 16/3, 419–40.

Mitchell, Tuti (1981). Hierarchy and Balance: a Study of Wanokaka Social Organization. Unpublished Ph.D. dissertation, Monash University, Melbourne.

Needham, Rodney (1983). *Sumba and the Slave Trade*, Working Paper No. 31, Melbourne: Monash University, Centre of Southeast Asian Studies.

—— (1987). *Counterpoints*. Berkeley, Los Angeles, and London: University of California Press.

Oembue, L. (n.d.) Fieldnote, Kodi, pp. 1–6. MS compiled by and received from Oembue Hina Kapita, Waingapu.

Reid, Anthony (1988). *Southeast Asia in the Age of Commerce 1450–1680: The Lands Below the Winds*. I. New Haven and London: Yale University Press.

Shaw, Isobel (1976). The Annual Horse *Pasola* in West Sumba. *The Indonesian Times*, 13 April, pp. 13–4.

Turner, Victor (1969). *The Ritual Process: Structure and Anti-Structure*. Chicago: Aldine.

Valeri, Valerio (1990). Diarchy and History on Hawaii and Tonga. In *Culture and History in the Pacific*, ed. Jukka Siikala, Helsinki: Finnish Academy Publications.

KATHRYN LOWRY

Between Speech and Song: Singing Contests at Northwest Chinese Festivals[1]

I

Several annual festivals in China, held primarily in areas which are
ethnically non-Han, were known before 1949 as temple fairs (*miao
hui*) or pilgrim fairs (*xiang hui*) and are currently characterised as
occasions for singing contests. Singing may resemble a conversation
between singers. Other songs 'quiz' an opponent on topics from
classical vernacular fiction such as the *Tale of Three Kingdoms*
which is known throughout China, on local history, or riddles on
on the weather, or pose frank sexual challenges or insults that may
demand an answer.

The regional songs called 'flower' (*hua-erh*) or 'young
man' (*shao-nian*) are primarily associated with popular religious
festivals prevalent over an area of some 100,000 square kilometers
in rural Northwest China, in Gansu, Qinghai, and Ningxia
Provinces (see map). They are amoebic verse, composed in
dialect, and sung to repetitive pentatonic melodies, usually
without instrumental accompaniment. Verses are alternated by
two or more men and women, who are commonly encircled by an
audience whose members may also decide to move into the contest
and sing. Participants in song exchanges are amateurs, who may or
may not be known as 'good singers'. There are professional
performers of flower song as well, most notably two members of
Provincial Song and Dance Troupes of Qinghai and Gansu, known
respectively as King and Queen of *Hua-erh*, who record these
mountain songs, sometimes with synthesised organ accompaniment.
However, professional singers do not 'contest' in song. They may
attend rural festivals to research folksongs which they revise and
adapt for stage performance and recording, but they do not enter
into song exchanges with the farmers, merchants, and workers
there.

What is the nature of a singing contest, and what is required for

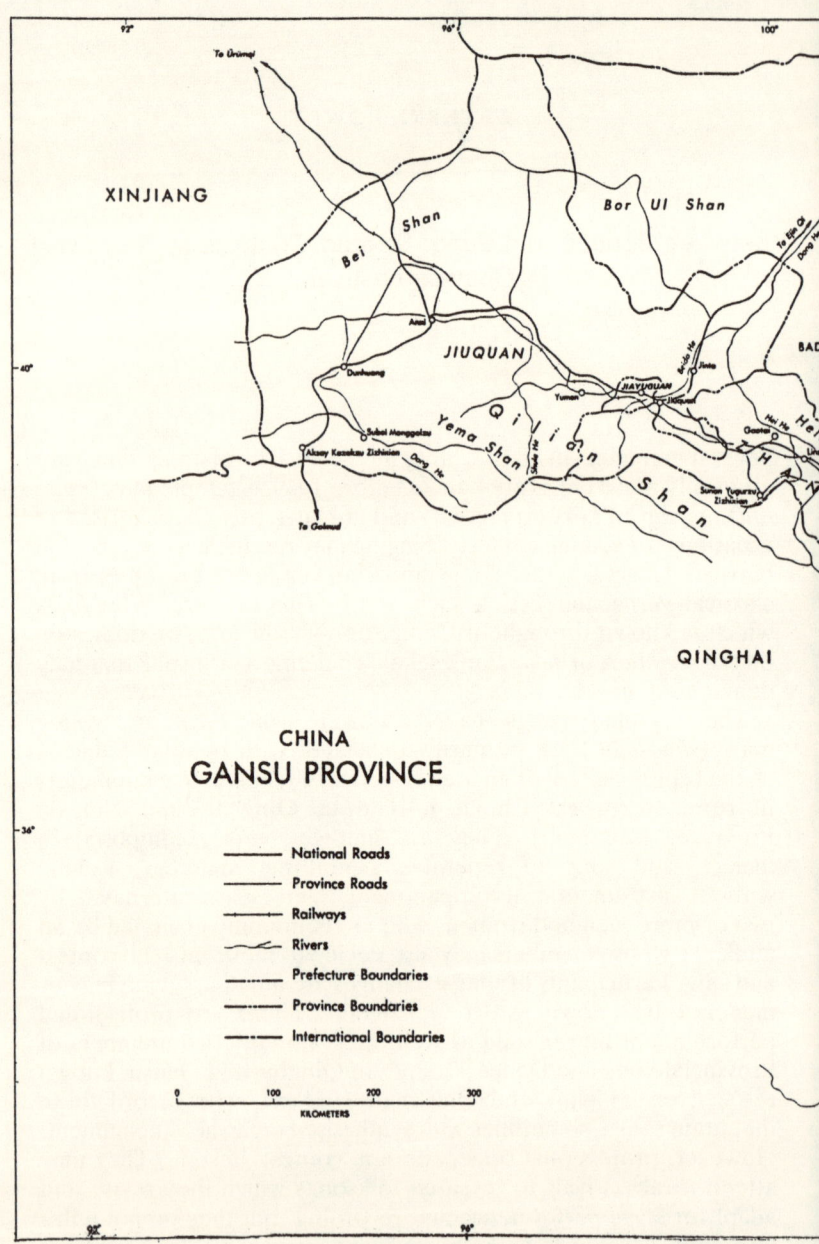

Map reproduced from *China: Growth and Development in Gansu Province* (The World Bank, 1988), showing location of Min County (Min Xian) Seat.

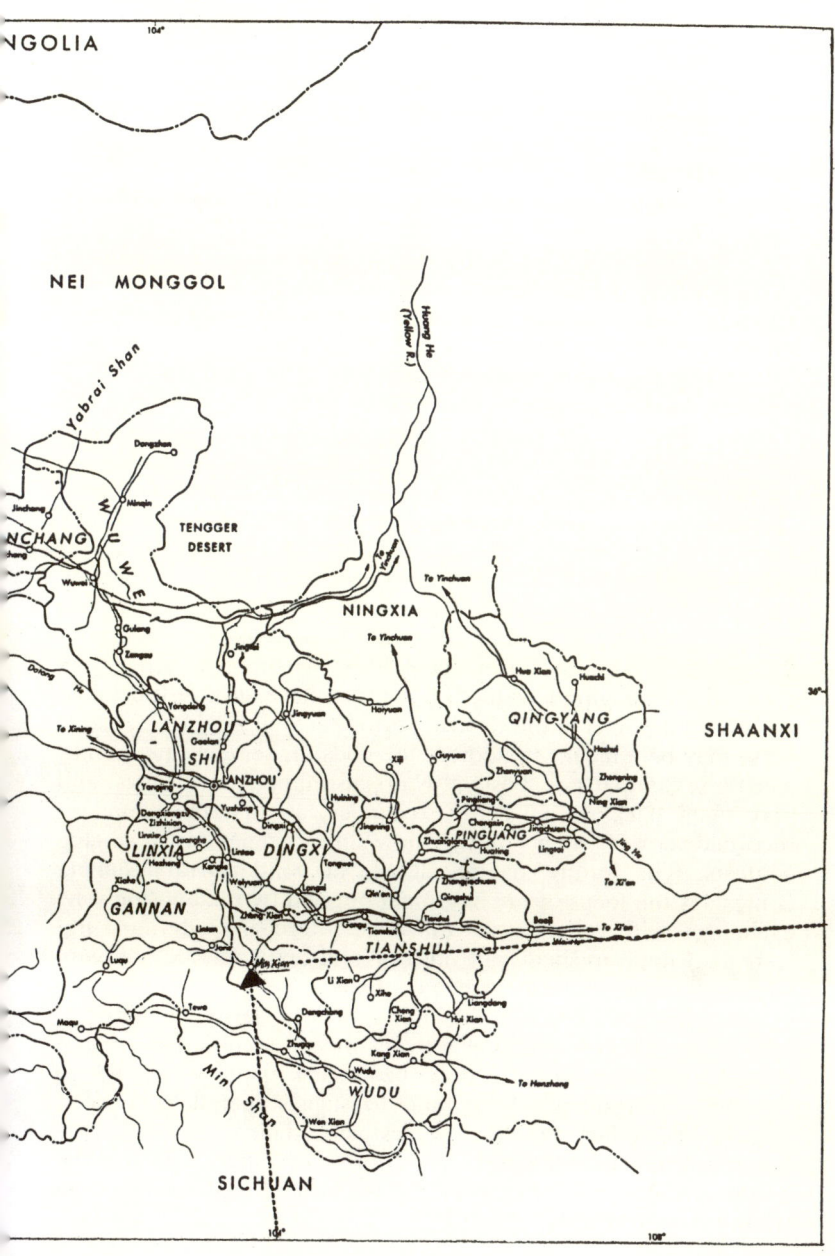

someone to become a participant? Why would someone want to
take part? If a song contest is presented as a win–lose situation, as
is commonly the case, can prestige or a prize be carried away
afterwards? Is it simply for the pleasure of singing that people
participate, or of humiliating someone when he or she falters or
fails to respond in song? For example, in response to a man who
confused one tune with another, and stopped mid-verse, a woman
sang:
>The 'flower' you learned you can't even sing.
>
>I bet that you can't even climb a ladder you've set up.

The interchange brought recording to a standstill, as audience and
singers burst out laughing (Lowry 1988: 11–3).[2]

I argue that song exchanges at Northwest Chinese festivals are a
peculiar kind of contest, one where most often there is no
'winner'. Rather, the primary criterion for being a good singer,
and accordingly for success in the contest, seems to be that of
continuity and variety of (oral) textual forms. In this respect, song
contests resemble spoken dialogue or conversation, in which any
member of the rural festival community can take part. This view of
singing is borne out by comments of native informants, and by
parallels to other festival activities. Yet interaction in singing
contests is not strictly verbal, nor can it be said to function strictly
as speech. In some performance contexts virtuousity rather than
continuity of singing is valued. Song may have several functions.
At different points in the festival period, as well as quizzing, sung
verse may be a means to address the gods, or serve as medium to
court a sexual partner. Festival-goers also tape-record songs, and
play them back during song exchanges, so that the sound of
recorded flower song 'music' now intermingles with singing
contests. The nature of interaction in contemporary singing
contests is the focus of my investigation, and by juxtaposing song
exchanges with other forms of social interchange during rural
festivals I hope to shed light on how oral texts operate and vary
with a given set of performance contexts.

Three sections in this writing intend, first, to clarify the nature of
interaction between singers at festivals by considering performance
contexts and terms used to describe singing matches. Second, to
examine the festival setting, and to suggest parallels between
singing and other activities at festivals. Third, to assess the
integrative function of song by considering narratives that treat the
origin of singing contests, and serve as charter for an ongoing song
tradition in Northwest China.

II

This section considers interaction between singers, as well as terms and criteria applied to performance. Many festivals are anniversary celebrations of the founding of temples to local deities and, since 1949, as noted above, are identified as the occasion for singing contests. In Gansu Province alone, as many as seventy-eight festivals are held between the first and the tenth months of the lunar year, each lasting one to four days. The majority take place during the fifth lunar month, that is in the lull between planting and harvest.

The Northwest festival songs are generally classified as 'mountain songs' (*shan ge*), a term indicating, first, that performance takes place outside the bounds of the community. Singing of *hua-er* tunes is generally restricted to festivals and to the time of weeding fields (Zhang 1986: 111). Second, verses are improvised to regional melodies, which are unmeasured and melismatic; the characteristic singing style uses high-pitched 'false voice'.

Min County, the site of festivals considered here, is in a high altitude region of dispersed agrarian villages punctuated by low mountain ridges. This area was once dominated by Tibetan and Mongolic peoples. Chinese scholars suggest that festivals date to the fourteenth century, and were established when Han Chinese began to establish settlements in the region (Ke Yang 1983: 88–90).[3] The population of southwest Gansu is composed of a large proportion of non-Han ethnic groups, many of whom are, or were, nomadic peoples. These include Han (Chinese), Hui (Chinese Muslim), Dong Xiang, Bao an, Sala, Tu, and Tibetan.[4]

In the festival context, the performance format for alternating songs can vary considerably (Pian 1988a; Lowry 1988).[5] In performance takes the form of an exchange that is most often referred to as *dui-ge* or *dui-chang*, a term I translate as 'song exchange'. In a recent study of the tradition of flower songs and scholarship, Sue Tuohy translates *dui-ge* as 'musical dialogue' (Tuohy 1988: 141 note). The term *dui* may mean to respond or reply, pair or couple, parallel or opposing, correct or proper, or face-to-face. All these meanings are facets of dialogue. Yet the character *dui* connotes a measured response, that matches an initial statement.[6] 'Exchange' better conveys the possibility that a verse may be faulted for not responding to another's in like coin, as well as the way in which successive verses of flower song are composed. The interaction between singers is more competitive than ordinary dialogue or conversation. Often successive verses are phrased as question and answer or challenge and response, and may be evaluated in terms of the appropriateness of response, or in terms of objective fact — usually based on fiction. Even when this is not the case, that is, when verses do not pose a direct question, a singer is expected to satisfy formal criteria, and may be faulted for errors in verse production. Criteria of end-rhyme, approximate quantitative metre, and coherence of theme and melody are mentioned only in cases when there is a 'mistake'. The 'best' singers are praised for breadth of repertoire (having 'a lot of lyrics in the stomach') and for vocal quality; the expression 'having a sweet throat' seems to refer to vocal control and ornamentation of a basic melodic line. It is somewhat misleading to call the exchange 'musical', as Tuohy suggests, since critical comments about singing primarily address features of verse, and singers and audience do not similarly pinpoint features of the conventional melodies.

Dui-ge may involve individual singers, or three to six singers, either all male or all female, who perform as a group, alternating lines within a verse. Regional tunes determine the differing performance formats, and a body of verse 'formulae'. Although the dynamic for the two forms of song exchange can be quite distinct, tunes seem to be used haphazardly. Essentially only two tunes are heard at festivals in Min County: one is called 'Piercing Knife' in Chinese, reportedly because of its shrill sound, pitched one or two octaves above the median range; in Tibetan, this tune is called '"Sweetheart" melody' (*Ah ou ling*) because it is commonly used for love lyrics. The second tune, '"Two Lotuses" melody', is sung by a group, and is highly repetitive. Each person sings a line or verse in turn, repeating the melodic line; then all unite to sing the refrain: 'Flower has two leaves'. A team of singers is made up of people from a single village, who seek out others who can sing at approximately the same level. The group includes one who composes and dictates lyrics to the others, called the 'joiner' or 'master joiner' (*chuan ba shih*). The *chuan ba shih* may also sing 'Piercing Knife' or other melodies, matching individually with others in song.

For example, successive verses in an exchange were sung:
(Three Women: to 'Two Lotuses' Tune)
1: [Ah] old man on the road [eh] take up a sickle to cut millet
2: [Ah] my love [ba ba] the one I long for hasn't come past
3: [Ah] night lamp shining [ma] intestines snapped heart
          broken
In unison: Ah flower has two leaves ah

(Woman: to 'Piercing Knife' Tune)
[Ah] on a distant [oh] way uneven, rocky white clay [yo] hill
[Ah] white peony thinking of you (I'm) sad so very [yo] sad
[Ah] white peony tears falling set the level mill grinding
Song exchanges are and are not rule governed. Verse production proceeds much as a game, without explicit abstract rules, but must

satisfy rigid implicit criteria. As noted above, singers must observe formal criteria of end-rhyme and approximate syllabic metre in verse production. Rhyme and metres are the most basic features; when these traits are observed, a verse 'flows from the mouth', *lang lang shang kou* (using the metaphor of waves). Further, conventional phrases are crafted to fit a given regional melody. 'Flower' song verse structure is governed by melodic repetitions, that most commonly result in a bipartite structure where topic and comment or metaphor and 'objective' statements are juxtaposed (Lowry 1985).

Within each circle, the order for singing is not clearly designated. At Min County festivals, several circles of singers alternated songs at the same time. Singers in different circles do not take heed of each other; song exchanges overlap, and in effect cover the area outside the central festival grounds.

The relationship of singers at festivals seems to be symmetrical, that is, anyone could conceivably take part in a song exchange, given the ability to compose or construct lyrics for the flower song melodies. There are, however, restrictions against alternating songs between family members and between people of different generations, especially across gender lines. Further, since the two melodies are regional, local to the southern part of the song festival area, ability to compose lyrics to a given tune and to ornament and vary lyrics for a long time is limited to the population of that area, as well as to singers who may be able to travel to a number of these events. People may sing when and for as long as they can improvise verses. In that sense, at the simplest level, participation is limited primarily by shared knowledge of the orally transmitted regional tunes, and corresponding verse style and themes.

III

Although annual festivals in Gansu are referred to as song festivals, singing is merely one aspect of festival activities. Further, it is literally a peripheral event, for most singing takes place on a hill or beside a river, in spots away from temporary shrines and opera performance. This section concerns the place of song contests in the festival and some of the parallels that contests evince to other activities.

Great Temple shoal *Da Miau tan* is one site in a string of ten or more festivals in Min County which take place over the course of nine days. Beginning on the 13th of the fifth lunar month, three festivals are held a few miles west of the administrative centre of Min County; a fourth is held on the 17th in the city itself, on and around a former temple site on Er-lang Mt; from there the festivals move to sites 8 miles north; then to two sites, 4 and 8 miles west along the Tau River, again to a site 15 miles north; then two festivals are held 15 miles west and 20 miles north on the 23rd day of the fifth month (see Appendix: Song festival sites). A diagram of the sequence of festivals describes an outward spiral, encompassing villages within a 25-mile radius along the Tau River.[7]

The *Da Miau tan* festival took place for the most part in the centre of the village of Xi Jai, which consists of about twenty households. This village was identified by local teachers and officials as Han Chinese, like others visited, but there were a number of merchants wearing the white caps which identify them as Chinese Muslim or *Hui*, as well as young Tibetan women in traditional costume.

The festival lasts for three days, as do most festivals in the circuit. For the duration, a temporary shrine is installed in the village elementary schoolroom, off a central courtyard (in a neighbouring village, a shrine is set up in the post office); an open stage for regional opera (*Qin qiang*) is located directly off the main road, diagonally across from the shrine; and beside it is a noodle shop. In contrast, song exchanges take place at a remove from other festival proceedings, on a hillside about a half a mile from the road through a potato field, beside the river, and in tents pitched in a grove of young poplar trees close by.

Each of the villages which holds a festival in Min County has a local god who has a double persona, both as an official whose deification was likely sanctioned by the government and as a rain god generally identified as Hei chih long wang (Dragon King of the Black Pool) or as a feminine deity Niang niang or Ah po believed to have power to grant fertility (Skinner 1964: 40).[8] During each festival the god is taken on tours of the village area, and is treated to continuous opera performances for much of the festival period; the deity's robes are changed, and he or she is given food and incense in the village shrine; there the deity also gives audience to prayers, generally for male children or for rain, and receives thanks for the harvest.

In past festivals in Min County, at some point in the proceedings, all ten gods would be placed in a square facing one another (Li Ling 1988). The significance of the 'congress' of deities is not entirely clear to me, but seems to represent a culmination of the festival. The convocation of gods with dual identities may be seen as analogous to the popular attendance at festivals. Positioning of gods within the square may be comparable to the positions singers take within circles of singer-opponents. Yet it is very difficult to conceive of Chinese gods competing, as that could only bring trouble to their devotees. If one follows this parallel through, the 'song exchange' (*dui ge*) may best be understood as 'song dialogue', a contest which draws people into a group rather than drawing distinctions between them.

Similarly, the continuity of opera performance for the duration of the festival represents a close parallel to song exchanges, whose aim is to continue to respond and vary the lyrics of conventional melodies. The performance of musical dramas is perhaps *the* characteristic element of annual festivals throughout China: staged for the duration of a festival, operas are meant to entertain the gods (Ward 1985; Tanaka 1985).

Popular deities are manifestations of the same god, who at the same time hold separate, secular personae. Yet local people do not dwell on the fact that gods have different personae, as Black Dragon god or feminine deity, and as deified officials. In addition to temporary shrines to these gods, there are a number of different sites and types of worship found side by side on the festival route: festivals may involve Buddhist worship and some chanting of *bao-juan* (vernacular Buddhist 'treasure texts'), elements of Taoist ritual offerings, and festival-goers continue to leave incense at the former sites of shrines to mountain gods,[9] which were destroyed in the 1958 campaign to 'Break the Four Olds'.[10]

IV

Several folk tales tell how singing contests became established as part of annual festivals. These narratives may shed light on the social significance of song exchanges. Some narratives depict singing contests as one form of several which might have been used to celebrate the founding of temples. Another pervasive concern that surfaces in tales is with the song name, and associations and symbolism of 'flower' (*hua-erh*). 'Flower' has sexual connotations, and may also indicate supernatural agency, of fairies or flower goddesses, in the origin of song.

'Flower' is a standard way of referring to a beautiful girl, one common in literary Chinese lyrics which has much older associations. It is also a term rich in sexual symbolism in popular religious practice in south central and other parts of China — an area where comparative research might show how the symbolism in songs and popular religion interrelate.

One way to describe a song exchange is: 'You sing Flower, and I will add leaves for you. Together we are a red flower with leaves, joined to make a fine bloom.'[11] This simile of adding leaves to

support or enhance the beauty of a flower describes the process of
alternating verse between two or more people into, and most
likely through, the night. The comparison between a flower with
its leaves and a woman matching with others in song recurs in folk
narratives about a flower fairy who first taught the regional songs
to people of the region.

Insight into two aspects of singing contests at festivals can be
drawn from the preceding simile. First, considerable value is
placed on breadth of lyrical repertoire. Perhaps the most common
criteria for a 'good' singer is whether he or she can continue to sing
when challenged and can complement or complete a theme
presented to him. When someone fails to respond to a sung
challenge, it is said to be socially humiliating.[12] Going back to
terms explained earlier, it is failure to be conversant. *Dui* is
commonly used in social language as well: the apology '*dui bu qi*' is
literally 'I cannot face you'. The singer who cannot answer
another's verse is sometimes said to have been 'defeated' (*shih bai
le*). As long as a singer can sustain an exchange, and keep coming
back with a response to others' challenges, he or she may continue
to sing for an entire day and night, or for the duration of the
festival. As noted above, the continuity of song exchanges
throughout the period of the festival presents a second parallel to
musical dramas which are staged continuously during festival time
to please the gods.

A second aspect of the song exchanges may further explain the
symbolism of the song name 'flower'. It is said that men and
women may reach an accord during a song exchange, and leave the
main festival grounds to tryst. Thus, song can be a medium to
identify a sexual partner, outside the bounds of the village and
away from the strictures of arranged marriage.

Another group of narratives concerns the origins of song
festivals, and provides a charter for singing contests. One story
tells how song contests came to be an established part of festivals
to celebrate the founding of local temples at Lotus Mountain:

> Long ago there was a beautiful blue lotus lake, where each
> year at the height of the summer the Goddess Mother of the
> West invited immortals to enjoy the scenery and to escape the
> heat. Not far from the lotus pool was a river (Yemu, a
> tributary of the Tau River that runs close to Lotus Mt), and in
> it was a black python who often left the river to wreak havoc
> in the surrounding area, and terrorise the people. One year,
> in the year of the dragon the Goddess of the West held a
> banquet of the Legendary Peach (*Pan tau hui*),[13] and she
> invited immortals and sages from all over the land to come.

Gold Flower Mother came from Mount Kunlun, seated on a cloud and carrying a thousand-year lotus bloom; as she was crossing Yemu River, the mountain valley was covered in fog, and then a line of black vapour rose toward the sky. Gold Flower Mother could see that it was a black python who was up to no good. She lightly set the lotus flower down, and suddenly there were ten thousand rays of coloured light. After that an earth-shaking sound was heard, and then a tall mountain in the form of a lotus was seen on the plain, located at the bend of the Ye Mu River, which sat on top of the trouble-making dragon. The people established the Jade Emperor Pavilion Yuhuang ge,[14] Zixiau gong, and Niangniang dian in that spot. At the celebration that followed, some were in favour of performing plays; some wanted to chant poetry; some wished to have horse-races. Because they were of different ethnic groups, and had differing customs, they could not agree on a single activity. Suddenly, just as they were all arguing, the sound of a prolonged, clear and sinuous melody could be heard from the sky, and then they saw two fairies. Each of these fairies held an umbrella in her left hand for shade, and in her right hand each lightly waved a coloured fan; they danced on top of the lotus peak that was encircled by clouds, and sang 'flower song' (*hua-erh*). People were enchanted by the songs, and could not help but harmonise to them. Lotus mountain was engulfed in a sea of song.

It is said that this day was the sixth day of the sixth month of the lunar year. Now every year from the first to the sixth of the sixth month is the Lotus Mountain Flower festival, when tens of thousands of people from the various counties of the autonomous region (including Lin-tan, Lin-tar, Weiyuan, Min and other counties). Further, singers of seven ethnic groups, including Han, Hui, Dongxiang, Sala, Tibetan and others, gather at Lotus mountain — they bring large umbrellas, wave bright coloured fans, and enjoy the landscape, giving themselves over to song dialogue (*dui-ge*).[15]

A composite of legends about 'flower' song singing and festivals makes ethnic differences in the beginning of singing contests explicit. Various printed versions of the tale say the Han wanted to perform plays; Tibetans wanted to read scriptures; and Muslims wanted to have horseback competitions.

Another legend about the origin of the festival at Gau Miaw, one of the festivals in Min County, tells that:

the Gods were pleased by the music they heard at religious fairs, and ordered the festival organizers (*shuitou*) through

oracles to give red cloths and gifts to good singers, so they
might sing mountain songs every year.[16]
In line with this myth, Chinese folklorists claim that local people
decorate those who sing well by giving them red sashes (*pei hong*).
I have never seen this done, but have seen people offer hard candy
or nuts, or bottles of locally manufactured orange sodapop, to
singers who are often stranded in the heat in the middle of a circle
of listeners for as long as they continue to sing. Again, when I
watched a song exchange in a public park involving two men, one
Han and one Hui, and a woman of Han nationality, people bought
several bottles of beer for them. The 'rewards' given to singers in
different contexts show how the differing treatments of the contest
may range from those officially sanctioned to purely secular, from
giving of sashes (a traditional way of honouring performers), to
giving sweets and nuts, to giving soda or beer to help favoured
singers continue to perform. There seems to be a common element
in the varying treatments of contest: as with the use of set
materials and forms in continuous song exchanges, rewards to
singers also prolong exchange within a group.

That singing contests are remembered in folk tales as one of
several alternative forms of contest to commemorate the founding
of temples is suggestive, and it raises the issue of the ethnic milieu
of singing contests. There is a pretence that flower song is a
common 'Chinese' (Han) expressive form agreed on by the
disparate cultural and linguistic groups in the Northwest. Yet, in
fact, singers may use Sino-Tibetan or Altaic phrases as well as
Chinese dialect. In addition, the population is ethnically and
religiously heterogeneous. Further, the nominal recasting of the
temple festivals as 'song festivals', noted above, places emphasis
on the secular element. Singing contests at annual festivals are
probably best understood as a part of multiple, contiguous local
festival traditions, which encompass a range of Taoist, Tibetan
Buddhist, and popular religious elements. Reasons that singing
contests are commonly represented as 'Chinese' and as secular
occasions cannot be pursued in this chapter.[17] The issue of the
function of singing contests for social integration will be discussed
in the concluding section of this chapter.

The flower song festivals, both the routes followed and the
attendance patterns, may be fruitfully examined in relation to the
rural marketing systems in China. In a series of articles on
marketing systems, Skinner termed the market towns 'culture-
bearing units', where social interaction of every kind is intensified
and broadened. People come together nominally to trade, but it is
in these places also that people seek entertainment, and seek the
help of a matchmaker.

Skinner distinguishes three types of market towns: standard, intermediate, and central. Though I will not go into a detailed summary of the marketing system here, an important point that he raises is the lack of correspondence between the marketing hierarchy and administrative system. Although all administrative centres also have markets, there is an essential difference between the administrative and the market systems. Markets form an interlocking network, and a standard market might have relationships with more than one intermediate market, and so on. In contrast, the administrative structure is hierarchical; subordinate to the administrative centre of a county there are x number of prefects, townships, and beneath that collectives (now villages), and households. Skinner argues persuasively that marketing 'had a significance for social integration in traditional China which both paralleled and surpassed . . . that of administration' (Skinner 1964: 31 ff.).

However, the temple fairs are yet again another network, and in certain areas of rural China may be a network which exhibits more cultural solidarity and stability than does the marketing system. Standard markets which are held several times a week (usually on alternate days three to five times in a ten- or twelve-day cycle) are not located in the same communities as the temple fairs. Markets and festivals are again two conflicting, or staggered, social structures, one organised on a regular cycle, the other on a lunar calendar.

Villages such as the one where the *Da Miao tan* festival took place are termed 'dispersed villages'. They may have a few shops but no market facilities, and are focused on a temple to a local god Tu di miao. In his study of *Religion in Chinese Society*, C.K. Yang writes of temple fairs that were intended to facilitate rural trade in a countryside where town and city economy were not sufficiently developed. While the economic function played the primary role, temple fairs were a community undertaking, and were the largest community gathering in which religion played an important role. Such fairs are particularly common in the North China countryside, especially in early spring and in winter, 'at a time that symbolized the beginning of another seasonal cycle of life and growth' (Yang 1961: 82).

In the preceding pages two apparently contrary claims have been made about singing contests. Emic terms about singing suggest it is a competitive activity, intended to humiliate those who lack verbal skill. Yet it was noted that if singing runs parallel to other aspects of the festival, it may best be viewed as dialogue and not as a competition.

The term *dui-ge* may best be understood as song dialogue in a broad sense: not as a fixed opposition between participants but rather as an open exchange between two to six or more singers. That dialogue occurs in overlapping groups, simultaneously, for the duration of the festival period. The several activities at festivals concurrent with song exchanges have been considered as parallels to the contests: popular deities are taken on a tour of villages within the temple circuit; operas are staged; people dress in new clothes, gossip, and mingle; goods — especially luxury goods — are laid out on the streets for sale. The collective community from an area 20–25 miles around most likely pools funds to pay for the opera; the area delineated by that collective undertaking is roughly the same as that of the community of singers.

Singing competitions differ from other festival activities primarily in the fluidity of form (oral songs, though implicitly rule-bound, are used in unrestricted exchange), and in the focus of songs on sexual themes. Conversely, there is extensive similarity between themes taken up in flower song 'question and answer' (wen-da) exchanges and the stories told in opera. Further, many of the tales and situations rehearsed in both media can be found throughout China.

The interaction during singing contests is double-edged: the songs may use bold sexual metaphors or simply refer to love relations; further, songs are in Chinese dialect which may incorporate phrases of Tibetan or (in other parts of the northwest) Altaic or Mongolic phrases. Both these traits mark singing as beyond the bounds of Confucian morality.[18] Further, song is physically outside of the village centre, and, as it occurs during festival time, outside the temporal frame of village activities. On the other hand, in singing contests people draw from a common repertoire of verse in dialect, and rehearse themes and tropes from vernacular fiction and history. Thus, song contests may create a unity of popular culture among the ethnically heterogeneous peoples of Northwest China.

Song contests, and the knowledge necessary to improvise songs and to take part, are regional phenomena. The forms and boundaries of festival contests, and areas in which various groupings cohere may be juxtaposed one with another. The description of singing contests given above sketches several spheres of activity that coincide in various ways with groupings who take part in oral song exchanges. Such an undertaking, carried out in the future in more detail, may give evidence of how verbal and musical contests can effect social solidarities which underlie administrative and other structures.

NOTES

1 Chinese terms are written in Pinyin romanisation. The present chapter is based on field research of regional songs and annual song festivals conducted in Gansu and Qinghai during 1985 and 1988. Research in 1985 led to a general study of song texts and regularities in orally composed song. A second period of research focused on song exchanges (*dui-ge*), here called singing contests, in the context of festivals mainly in southwestern Gansu Province, Min County (*Min Xian*, see map). Research was supported by grants from the Jacob K. Javits Foundation, U.S. Department of Education (1988), and joint support of Exxon/Woodrow Wilson School, Princeton University (1984). I am grateful to Professors Stephen Owen and James L. Watson for their comments on drafts of this report. Errors are my own.

2 Exchange between a man, Yang Zhongchao (50-years old) and woman, Luguo Guancai (55-years old) recorded in Qingshui Village, meeting room of the village government. The two were said to perform frequently against one another.

3 There is debate over whether festivals first were held in the Ming dynasty Hong wu (1368–98) or Jia qing (1522–45) reigns. In both periods Chinese garrison towns were kept in this area, considered 'barbarian' territory since well before the previous Yuan dynasty (1279–1368).

4 The *General Survey of Linxia Hui Autonomous Region* from 1986 records the population of this area includes Han Chinese 47.2 per cent; Hui (Chinese Muslim) 35 per cent; Dong Xiang 16.8 per cent; Bao An 0.58 per cent; Sala 0.3 per cent, Tu and Tibetans.

5 Lotus Mountain festival described by Pian takes place roughly 50 miles northwest of Min County, and begins ten days later than the end of the festivals in Min County.

6 The definition of the archaic character *dui* in the Han dynasty (212 BC–AD 220) dictionary *Shuo Wen* uses the figure of strokes on a bell, which are measured in terms of force and volume to respond (and correspond) to an initial stroke.

7 The route of the largest annual flower song festival in Gansu, where an estimated 50 000 people attend, starts at Lotus (*Lian Hua*) Mountain, and similarly moves through five sites in a radius of 20 miles over a six-day period. There are a number of designated sites on the mountain, a series of gates and former shrines to mountain gods and popular deities, where people continue to place incense, though the structures were destroyed in 1958 and 1968.

8 Skinner cites from a study of temples and folklore in

Northern China that a cult of the Black Dragon Hei Long is concentrated in a single section of Wangquan County, and other maps suggest the six temples to that deity coincide with a standard marketing area. This cult may or may not be related to the one I describe for Min County (see Grootaers 1948).

9 Offerings to mountain spirits are a type of worship said to be related to Tibetan practice. Gazetteers note shamanistic practices in this area of the Northwest in the late eighteenth century (Qing dynasty, Qian long reign).

10 The campaign against 'old customs, old habits, old ideology, and old culture' was part of the Cultural Revolution (1968–78) in the PRC.

11 This is a common description of interaction between singers of 'Two Lotuses' tune, where each singer sings a line in turn, and then sing the refrain in unison: 'Flower has two leaves'. The quote was used in an article about the Queen of Flower song Su Ping (Ma Xiaoxiao 1984: 29).

12 This social humiliation is not 'loss of face', but a mild embarrassment. The singer who cannot answer a challenge is *bu hao yi si*. Interview, Zhu Gang and Ye Jinyuan, 1984.

13 *Pan tau*, flat peach or saucer peach, enables one who eats it to live forever. This feast is held on the 3rd day of the 3rd lunar month, to honour the Goddess Mother of the West, Xi tian Wang mu.

14 *Yuhuang* or *Yuhuang da di* is the supreme Taoist deity.

15 This account, in *General Survey* (1986: 26–7) is similar to other tales about festival origins. Tuohy notes of composite myths (below) other versions which say Tibetans wanted to have songs and dances, and Muslims wanted to chant scripture. See Tuohy 1988: 212, 244 note. See also version in *Lotus Mountain* (1980), magazine of Kang-lo County Cultural Bureau, Gansu, pp. 24–5.

16 Tuohy (1988: 212) translates a report by Chen Ming (1985).

17 See Tuohy 1988 for a thorough and insightful treatment of scholarship on Flower song festivals, and of issues that have shaped research and the festivals themselves.

18 See Chang-tai Hung (1985: 65–9, 75 ff.) on literati interest in collecting love songs and obscene folksongs early in this century.

REFERENCES

China: Growth and Development in Gansu Province (1988). Washington, DC: The World Bank.

General Survey of Linxia Hui Autonomous Region (1986). Gansu Minorities Press.

Grootaers, Willem A. (1948). Temples and History of Wanch'uan (chatai), the Geographical Method Applied in Folklore. *Monumenta Serica* XIII, 209–16.

Hung, Chang-tai (1985). *Going to the People: Chinese Intellectuals and Folk Literature*. Cambridge, MA.: Harvard University Press.

Ke, Yang (1983). *Hua-er di yuan-liu* [Sources of *Hua-er*]. In *Hua-er lun-ji* [*Collected Essays on* Hua-er] I. ed. Gansu Division of Chinese Folk Literature and Arts Research Society, pp. 86–105. Gansu Peoples Press.

Li, Ling (July 1988). Interview concerning unpublished research on festivals, Min County, Gansu.

Lowry, Kathryn (1985). Language, Music and Ritual: Flower Songs (Hua-erh) of Northwest China. Unpublished B.A. thesis. Princeton University.

—— (1988). Change and the tradition of flower songs and festivals. Unpublished paper.

Ma, Xiao-xiao (1984). '*Hua-er*' *ge-chang jia* [A 'Hua-erh' singer]. *Xin Guancha [New Observations]* 6, 28–31.

Pian, Rulan Chao (1988a). Flower Songs of Lotus Mountain: an Investigation of Context. Unpublished draft.

—— (1988b). *Cong Hua-erh di yan-chang tan dao biaoyan shih-kuang wen-ti* [Comments on the issue of performance context based on flower song performance]. Draft.

Skinner, G. William (1964). Marketing and Social Structures in Rural China, Pt I. In *Journal of Asian Studies* 24, 3–43.

Tanaka, Issei (1985). The Social and Historical Context of Ming-Ch'ing Local Drama. In *Popular Culture in Late Imperial China*, eds David Johnson, Andrew Nathan, and Evelyn Rawski, pp. 143–60. Berkeley: University of California Press.

Tuohy, Sue (1988). Imagining the Chinese Tradition: the Case of *Hua'er* Songs, Festivals, and Scholarship. Unpublished Ph.D. Thesis, Indiana University Department of Folklore.

Ward, Barbara (1985). Regional Operas and their Audiences: Evidence from Hong Kong. In *Popular Culture in Late Imperial China*, eds David Johnson, Andrew Nathan, and Evelyn Rawski, pp. 161–87. Berkeley: University of California Press.

Yang, C. K. (1961). *Religion in Chinese Society: A Study of Contemporary Social Functions of Religion and Some of their Historical Functions*. Berkeley: University of California Press.

Zhang, Yaxiong (1986). *Hua-er ji* [*Collected* Hua-er]. Revised edition. Beijing: China Arts Press.

Zhu, Gang and Jinyauan, Ye (Interviews 1984 and 1988).

Qinghai Minorities Institute, Chinese Department.
Xining, Qinghai.

APPENDIX: SONG FESTIVAL SITES, MIN COUNTY,
GANSU, PRC.

Dates refer to the lunar calendar. The single date noted is the
festival peak; start and end of the festival noted in paren-
theses. The first festivals in this sequence occur near Min
County Seat, *Min Xian cheng*, about 3 miles west of the City,
and move from there east into the city and then both west and
north as much as 20 miles along the Tau River.

1 *Da gou jai* festival. 5th month, 14th (13–15). 50,000
  people.
  Takes place three and a half miles west of Min County
  Seat, on the north side of Tau River. Held on a mountain
  which was originally full forest, but was burned down in
  the 1960s; this event is said to have become a song festival
  only recently.

2 *Miao-erh tan.* 5th month, 15th day (13–15).
  1 mile west of Min County Seat, south side of Tau River.

3 *Ma-shih tou.* 5th month, 15th day.
  2 miles southwest of Min County Seat. The festival takes
  place on rocks to the north of the Tau River, in or next to
  'Horse Rock' village. There is a monastery very close to
  this, Clifftop Monastery (Shang-yen si) on Min Mountain.

4 Erh-lang Mountain. 5th month, 17th day (15–17).
  On the south edge of Min County Seat, south of the Tau
  River.

5 *Gau Miao.* 5th month, 20th day.
  The festival is held at Mei Chuan, also known as 'Base of
  the Shrine', 8 miles north of Min County Seat, and ½ mile
  east of the Tau River. It is a temple festival, not a flower
  song festival, but people may sing there now.

6 *Gan Jai cun.* 5th month, 20th day.
  This village is 4 miles west of Qinqshui south side of Tau
  River (see 7).

7 *Erh-shih he tan.* 5th month, 21st day (20–22).
  Festival held in Qingshui Village, 8 miles west of Min
  County Seat, north side of Tau River. The peak of festival
  singing occurs at night on the bridge over Crowtooth
  Ravine (*Ya-wu Gou*).

8 *Gu cheng.* 5th month, 22nd day (21–22).
  This village is 15–16 miles north of Min County Seat.
  Festival takes place beside a dam, east of the Tau River.
  There may also be a second festival here on the 6–8th of
  the 6th lunar month.

9 *Da-miao tan.* 5th month, 23rd day (22–24).
  The 'Great Temple shoal' festival held in Xi Jai village

approximately 15 miles west of Min County Seat, north along Tau River.

10 *Mei-chang tan.* 5th month, 23rd day (22–24).
In Zhong Jai village, 20 miles north of Min County Seat, on the north side of Tau River.

JOY HENDRY

Children's Contests in Japan

I THE NOTION OF COMPETITION IN JAPAN

It may come as a surprise to some to discover that the Japanese people, whom we have come to regard as fierce rivals and competitors in the world of big business, and whose school-children, we are told, fight with great dedication to achieve the highest standard in educational accomplishment, in fact only adopted a word which approximately corresponds to the Western notion of 'contest' or 'competition' in the late nineteenth century. This word, kyōsō (競争) was created, along with a number of others, by Fukuzawa Yukichi (de Mente 1983: 59), a man who dedicated much of his life to understanding, introducing, and interpreting Western notions to his compatriots, out of Chinese characters which were already in use in other ways.[1]

In the case of this particular word, two characters have been used. Both are employed independently as verbs, the first (競) often translated now as 'to compete', although its English dictionary entries also include 'to compare' and 'to emulate'; and the second(争) to dispute', 'to argue', and 'to be at variance with' (Nelson 1962; Kenkyusha's *New Japanese–English Dictionary*). It is generally dangerous to speculate about the internal make-up of characters themselves, but in a language where something of the feeling for a concept is derived from the look of the character which represents it, it is perhaps worth noting that the first of these two characters consists of two identical halves, each made up of the character for brother, or, more precisely, elder brother, with the character for 'to stand', which includes many of the same connotations as its counterpart in English, set above it.

This brief analysis of kyōsō does already suggest one or two reasons why the Western concept was difficult in Japanese. There is, for example, no notion of 'brother', with any implication of equality. One is always an elder or a younger brother, and this disparity is reproduced in most other relationships in Japanese society. A person is surrounded not by equals, with whom he or she can compete on the same terms in various arenas,[2] but by people who literally 'go before' or 'come after', with consequent implications of appropriate behaviour. The traditional form of learning was by the emulation of an elder, or superior in this sense of 'going before', and this superior/inferior dyad was fixed permanently so that competition between pupil and teacher was quite inappropriate. In practice, of course, comparisons would be inevitable, and there is a veritable battery of social constructs which aim to maintain harmony between people in positions of potential rivalry.

These social constructs include the emphasis in Japanese socialisation on co-operation and compromise, and much of the training of small children in Japan is dedicated to inculcating the value of these notions and various mechanisms for achieving them.[3] First, with regard to co-operation, individuals are expected to put their own interests second to those of wider groups to which they belong, and effort is dedicated not so much towards personal prowess as to prowess which benefits this group. In practice, as will shortly be demonstrated, competition is perfectly acceptable between groups, indeed part of the point of group co-operation is, at least nowadays, in the pursuit of competition between groups.

Putting this point together with the notion which shares emulation with comparison and competition may help our understanding of Japan's success in the world of modern business, as well as our understanding of the indigenous concept. For many years after the Meiji Restoration, when Japan opened its doors following two hundred years of seclusion to Western commerce and communication, 'the West' — in practice, first Europe, later the USA — was seen as ahead, in the sense of 'going before' and therefore superior in Japanese terms. Learning through emulation was quite appropriate, therefore. But since the Western world remained composed of outside groups, as far as Japan was concerned, it was also perfectly legitimate eventually to see these groups as rivals with whom to compete and ultimately defeat in one way or another.

As to the importance of compromise, this is well illustrated by a basic principle of judgement in Japanese courts. Here, it has been often pointed out, the aim is not to establish who is right and who is wrong in any particular case, but rather to achieve a settlement between the disputants which both accept as fair.[4] In other words, there is less attempt to judge, although the Western-inspired legal system includes such a role, than to mediate between people who often enough accept that they are both to some extent at fault to have got themselves into a dispute in the first place. As de Mente notes in his entry on 'competition' in *The Whole Japan Book* (1983: 59), 'the primary Japanese principle has always been co-operation — try to get along, by compromising, so that no one loses and all gain'.

II CHILDREN'S CONTESTS: SOME EXAMPLES

Despite this considerable ideology apparently opposing competition in Japan, there is no doubt whatsoever that children and adults in Japan today, as in the near and distant past, engage in a wide variety of activities which may be classed as 'contests'. Some examples of these have even become extremely popular in Western countries. Martial arts, many of them Japanese, are now readily available as a form of sport or exercise in most parts of the Western world, and *sumo* wrestling has recently become a popular spectator sport, shown regularly at prime viewing time on British television. Moreover, perhaps paradoxically in view of the emphasis on group co-operation, there are no team sports indigenous to Japan, although several Western ones have been adopted and adapted there. Let us turn to an examination of some examples of children's contests to see if it will help to us decipher this seeming paradox.

To start with an extremely modern occupation, small children in the Japan of the 1980s spend a great deal of time in front of a television screen, not necessarily watching television, although this, of course, happens too, but very often engaging in the numerous electronic contests devised for them by computer companies. Known as *famicon*, a Japanese contraction of 'family computer', these machines are virtually ubiquitous in Japanese homes with children, and the games they offer have a variety of Western-sounding names exemplified by the series entitled Mario-man, Mario Bros, and Super-Mario. These games may be played alone, when a child is generally engaged in improving his or her own skill, and perhaps beating a previous record, but they usually also have two control sets so that they may be played as a contest. Japanese versions of these games are purpose-built and relatively cheap to acquire.

To engage in contest in this way is by no means exclusively a modern, Western invention, however. Many more traditional children's games could also be played alone, or in competition with others. In the early part of the year, it has for long been customary for children in Japan to fly kites, often splendid new ones received as presents at celebrations for the New Year. Again, one could spend time practising the various movements one should make to allow the wind to manipulate one's kite, but it is also an old custom in some provinces purposely to fight with kites in the air 'until the string of one is snapped to the discomfiture of its owner' (Iwado 1936: 24). These kites are invariably decorated in bright colours, and many of them, given particularly to boys, depict heroic figures of Japanese military history. Iwado (1936: 23–4) supposes that this is due to the fact that kite-flying was a pastime encouraged in the feudal days of the Tokugawa period as a sport becoming the sons of samurai families, when these military heroes were to provide models for emulation.

Popular New Year gifts to girls are also symbols of contest, although somewhat more subtle in their meaning. They are *hagoita*, usually translated as battledores, since they are ostensibly to play with shuttlecocks, but the backs of them are painted or elaborately decorated with cloth and padding to create designs sometimes so delicate and beautiful that they have become *objets d'art*. In fact, they are probably more often displayed in glass cases than actually used in games, although this elaboration was apparently a development of the Tokugawa period (1600–1868) and they were recorded as used in games at court (by boys) as long ago as 1432 (*Kodansha Encyclopaedia*, III; *Sekai Daihakka Jiten*). Perhaps since then girls engage in contests with one another as much to show off the beauty of their battledores as to win points over one another in the striking of the shuttlecock. There is also an element of ritual involved here, as the gifts are given in some areas to celebrate the birth of a new baby girl in a family, and this aspect may have some significance, as will shortly be discussed.

Another game traditionally popular among Japanese children is top-spinning, which also involves an element of contest. The tops are made in a variety of shapes, of a variety of materials, but they are usually spun by means of pulling sharply on a knotted rope which is first wound around the top. The game is played on a limited, usually concave surface, such as a cushion, or a piece of cloth or plastic stretched over a bucket or barrel, and the idea is to spin one's own top in such a way that it knocks the top of the opponent out of the space. There is something vaguely reminiscent of the *sumo* ring here, since this is also a contest involving the dislodging of an opponent from a limited space, but we will return to the influence of *sumo* on the Japanese notion of contest shortly.

According to T. W. Johnson, who carried out a study of adolescent peer groups in Japan, the Ministry of Education disapproves of games such as these. Another example is again a kind of contest where children throw cards down, creating a breeze to flip over the cards of their opponents. Johnson records that these games were prohibited within school premises, partly because they are classified as forms of gambling, which is viewed very negatively. Teachers are expected to confiscate tops and cards, which often come free with candy and other purchases children are wont to make, whenever they catch children with them (1975: 226). There is also an ideological aversion for various forms of individual competition within schools, and this is made especially clear in kindergartens.

The sports day, for example, which is perhaps the epitome of competitive activity in Western schools, is a highly valued event in Japan, too, but it is not individual competition which is emphasised. Instead, races are run so that pupils may use their ability to benefit wider groups, or to co-operate with one another. Several races are usually held between teams which represent the catchment areas from which the school draws its clientele, and each area chooses the children who are known to be best at the particular events in question. Events like three-legged races are particularly favoured, although they may be five-legged, or even nine-legged, so skilful become these children at moving in concert with one another, and the tug-of-war is the highlight of the day. Here there is real opportunity to pull together with the group, or bite the dust with them, and the line of children engaged in the duel is usually matched by a line of parents recording on film their offsprings' participation in the ultimate co-operative venture. The entire school, or kindergarten, is divided into red and white teams — a compulsory part of the uniform is reversible caps — and at the end of the day, an overall win is announced for red or white.

Informally, of course, children need to know who is the most skilful at any particular activity so that they can choose the best representatives for the groups to which they belong, and these skills are determined by holding races between individuals in the weeks preceding the main event. There is undoubtedly considerable competition between possible candidates, and this may lead to unpleasantness behind the scenes. It is possibly partly for this reason that competition within groups is discouraged so strongly, for it is the ideal that members of the same group will preserve harmony amongst themselves. In class, too, children are not given official class places as part of their examination results, rather they officially measure their progress against previous achievement, although they usually find out from their friends approximately how they stand.

Other forms of children's contest are actually encouraged by adults for their educational value, however. These include versions of the card game known as *karuta*, which involves matching up two halves of something. In an ancient form, the two halves were clam shells, which are apparently never alike enough to fit with any other than the original partner, but a traditional version of the card game, again associated particularly with New Year, is *uta karuta*, where well-known poems are written in two parts. The first and second parts are then separated into two decks of one hundred cards each. The deck containing the second halves is divided between the two sides taking part and laid down. The other deck is read out, card by card, and the two sides compete to find the remaining lines of the poem, if necessary by snatching the appropriate card from the other side. This appropriated card is then exchanged for two of their own, the victor being the side which first disposes of all its cards. Clearly, the team whose members are better versed in poetry will be sooner able to locate the cards which complete the snatches read out, and this therefore encourages the participants to memorise their classical poetry.

Sets of *karuta* also come with large clear characters, to be matched up with the sounds they make, or with pictures of objects beginning with those sounds, and these are given to small children learning to read. Others have popular sayings or well-known phrases on them. One version of the game, issued free by the local education authority to all children in the Yame district of central Kyushu, has various phrases exemplifying the local dialect of the that district. The officer in charge of the project explained that much of the local language is disappearing with the standardisation of Japanese encouraged by school and television, so they were distributing the game to foster awareness and knowledge of local traditional culture through a medium they knew would be popular. In this case the fact that it is played as a contest seems to bother the education authorities not a bit.

Evidently, contests may be viewed positively if they have some further aim, for the most common form of contest among Japanese children may quite acceptably be played innumerable times in any single day. This is a game familiar in the West, too, where the contestants shake their fists twice before deciding whether to form them into one of three possible shapes, known in English as stone, scissors, and paper, in Japanese as *gu, chokki, pa*.[5] The participant who forms stone defeats the one who forms scissors, as stone will blunt scissors; the scissors defeat paper, however, as scissors will cut paper; and paper defeats stone because paper can wrap stone. This game can also be played with the feet, where the participants jump into one of the three positions that also signify stone, scissors, and paper.

The game can be played purely as a contest, and among young children it may be played with some kind of a forfeit for the loser, one source suggesting that children of seven or eight will often be required to show their genitals as a forfeit (Johnson 1975: 232). It is much more commonly played as a means of making decisions, however, or even as a way of resolving disputes. There is little skill involved, except possibly that required to anticipate the thoughts of one's opponents, and the outcome of such a contest usually resolves a difference of opinion once and for all. It can be a means to decide who goes first in another game, or who goes out to buy the newspaper. It can be a way of amusing children as well as helping to avoid disputes among them. It can also be a way of dividing a group of people into teams, or several smaller groups.

Another way of doing the latter is to have the participants wave their hands in the air while chanting *omote-ura*, or 'face' and 'reverse', and then at a predetermined moment choosing to expose either the palm or the back of the hand simultaneously. Teams may be selected in this totally random way, and there are further permutations for evening out the numbers should the group fail to fall into two equal parts. In fact, Japanese children know a large number of variants on both these decision-making contests, and they move through the selection appropriate to a particular situation with a speed quite alarming to an outsider naively trying to participate. Even if one understands the rules, it probably takes some considerable practice for an adult to be able to participate smoothly and fully in a complicated decision-making process of this kind.

III SOME GENERAL CHARACTERISTICS OF JAPANESE CONTEST

The element of chance involved in this activity is, in fact, possibly one of the most important distinguishing characteristics of Japanese contests in general. It is said that, in the past, contests held at Shinto shrines were often carried out to determine the wishes of the god or deity of that particular location. Horse racing, archery contests, tugs-of-war, and wrestling matches were apparently all held as a means of allowing some transcendental being to foretell the future. Depending on the victory, or defeat, the contests were seen as a kind of oracle indicating to farmers information about their harvests, or to fishermen about their future catches (Yoshida *et al.* 1987: 8, 40). This ritual aspect of contests is still evident in many of the martial arts and particularly in the *sumo* contest, which is heavily imbued with Shinto ritual. A brief examination of the latter is quite a revealing exercise.

According to a legend recorded in the earliest Japanese chronicles, the original unity of the Japanese people was established by a *sumo* match, when the god, Takemikazuchi, won a bout with the leader of a rival tribe.[6] This ancient sport would seem to date back at least some 1500 years, when it was principally a ritual dedicated to the gods, together with sacred music and dancing, to propitiate them or to thank them for a good harvest or other bounty they were seen to influence. In later periods it became a regular courtly activity, too, but the court was also hedged about with sacred qualities, and the contest is still highly formalised, with clearly sacred elements. During the opening ceremony, for example, the higher ranking participants wear a large twisted rope around their waists, a *shimenawa* like the straw rope used widely in shrines and elsewhere as a marker of the sacred. They also throw salt in the ring before they start, as a ritual of purification.

The bout itself actually forms only the climax to a great deal of physical, ritual, and psychological preparation, but it is the outcome of the bout which ultimately determines the ranking of the wrestlers who participate professionally in the sport. The aim is to force the opponent either out of an inner circle a little over 15 ft in diameter or to cause him to touch the ground with any part of his body other than the soles of his feet. Various rules of procedure are to be observed, and these are rigorously enforced by a referee, who looks virtually indistinguishable from a Shinto priest and calls out in a high-pitched voice throughout the contest, and by a team of five judges, who sit in the first row of the audience. There are six grand national tournaments per year in Japan and each continues for fifteen days. They are broadcast on television, and the live audience is usually big, although much of the actual time involved is spent in ceremony and in a pre-bout preparation period. This includes a session of 'cold warfare', recently restricted to four minutes, when the opponents adopt a fierce posture and glare at one another.

Children's *sumo* has been popular at various periods in history. It has often been a feature of shrine festivals, for example, and may have once been a part of fertility rites (Cuyler 1985: 29–30). In the area of Kyushu where I carried out fieldwork[7] it was stopped within living memory because of an injury, but elsewhere it is said still to be held. Many festive activities involve children and youths of an area, who are traditionally held to be closer than adults to the gods. There is also considerable entertainment value for the wider public in such an event, although the children themselves undoubtedly take the competitive side seriously. During warring periods of Japanese history, particularly between the fourteenth and sixteenth centuries, *sumo* was principally regarded as a martial art, indeed, it developed in this form into *jujitsu* (Cuyler 1985: 46), the unarmed combat which later became known as *judo* because this latter word emphasises the spiritual side of the skill rather than the military one.

Currently, the word *sumo* does, in fact, carry something of the meaning of 'contest' in itself, for it is used in a couple of other children's activities. The first is rather familiar in the West, too, where two participants pit the strength of their forearms against each other, pivoting themselves at the elbow to try and knock over the raised arm of the other. In Japan, this is called *udezumo* (腕相撲), where *ude* (腕) is the Japanese word for 'elbow', and *zumo* (相撲) simply a softened version of *sumo*. A similar kind of contest is *yubizumo* (指相撲), where the strength of the thumbs is at stake. The similarity between top-spinning contests and the rules of *sumo* has already been noted earlier, and there is another contest popular with children which is somewhat reminiscent of *sumo*. This is *niramekko*. In this, the idea is to fix an opponent with a stare, although it does not matter if you blink, and the victor is the one who can refrain for longest from laughing or from showing his or her teeth. The contestants clamp their mouths shut and proceed to distort their faces as wildly as possible in the hope of disrupting the composure of their opponent into laughter.

The Chinese characters for *sumo* are quite straightforwardly a combination of one (相) which has the meaning of 'mutually', 'reciprocally', or 'together', and another (撲) which means to 'strike', 'beat', or 'knock down'. Until around the tenth century, however, *sumo* was apparently pronounced *sumai*, and often written with a second character meaning 'dance'. In fact, the competitive aspect of *sumo* wrestling has now almost completely superseded its religious meaning for most of its afficionados, although some of the ritual is maintained, possibly to sweeten the proclivity to gambling which it inspires.[8] The wrestlers are also professionals, dedicated completely to the pursuit of the sport, or art, and they spend the better part of their lives training for the almost transitory bouts which constitute the contest.

According to Yoshida *et al.* (1987: 9), this transition from the ritual significance of competitions to a secular, sporting one, is a historical development common to most contests in Japan. Victory or defeat is no longer an oracle signifying the will of the gods, but is a demonstration of personal and physical prowess. This leads to intense discipline and training to improve these skills, and an emphasis on perseverance and effort in Japan has perhaps made the pursuit of skill an activity almost as important, if not sometimes more so, than the contest itself. This principle could certainly be applied to martial arts, which may well have aims beyond the development of skills, but it is also illustrated in another example from children's activities, namely Johnson's description of Sunday baseball played by the youth group he studied.

According to this report, the game starts as soon as it is light enough to see the ball, and it continues until the ball becomes invisible again in the evening. Boys come and go during this time and may join either of the two teams engaged in play. No one keeps score, and Johnson was himself reprimanded for trying to do so, although when he surreptitiously continued he discovered that there was very little between the teams throughout the whole day. This situation was achieved quite consciously, apparently, by children changing sides if one became stronger than the other, and the aim of the whole activity was evidently to practise and to enjoy the day's play rather than to win or lose a match. On occasions, however, matches would be arranged against teams from other towns, when the best players would be selected and the aim was very definitely to achieve victory. Similar principles applied to the winter activity of football (1975: 137–41).

This proclivity in Japanese versions of even Western sports to play for play's sake rather than to enter into a contest is rather well known amongst Westerners who have arranged to play various games with their Japanese friends. To take tennis as another example, Japanese players will happily spend a whole afternoon 'knocking up' in our terms, with no attempt to arrange a game between the participants. Even if 'games' are set up, these may well involve the rotation of the people involved amongst different possible partners, so that there is no overall winner or loser at the end of the day. Comments may well be made about the skills of individual players, but efforts will be made to counteract the overall advantage of the most adept by pairing them with the least able. In this way, competition is indeed eliminated as the primary target of the activity, at least when members of the same inside group play together.

IV CONCLUSION

This elimination of competition among members of the same inside group is undoubtedly the crux of the matter. As mentioned at the beginning of the chapter, contest and competition should be aimed out of groups, the relationships among the members of which are geared to maintaining harmony. The skills developed may, of course, be demonstrated in competitive arenas, but these will be against members of other possibly parallel groups rather than against one's own associates. In schools and kindergartens, children may well find their friends in opposing teams, but the boundaries of their allegiances will be clearly based on some objective criteria such as residence or formroom. At this stage somewhat artificial boundaries may need to be established while the children are absorbing the basic principles of directing competition out of the group.

I have already supplied ample evidence that children engage spontaneously in any number of contests amongst themselves, with their friends, preferably unsupervised by adults, and I have suggested that these are important ways in which they come to assess one another. To explain this apparent paradox, I suggest that it is concerned with another important distinction in Japanese society, namely that between *omote* and *ura*, or 'front' and 'rear', behaviour. To engage in contest is one of the ways in which members of a group come to know one another, if necessary to select one another for official competition between groups, but this revelation of strength and weakness within a group should be strictly reserved for behind the scenes, the *ura* (rear) area.

In an official, up-front world, usually that supervised by adults such as representatives of the Ministry of Education, there is a certain prevailing ideology about competition and its appropriateness or inappropriateness to particular occasions. The strong emphasis on harmony within co-operation groups makes it important to minimise overt aggression within these groups, although there is plenty of evidence to suggest that the harmony is actually a rather carefully manicured veneer.[9] Children are in general permitted to acquire their social skills gradually in Japanese society, and they are not often punished for activities which do not interfere overtly with the others imposed on their lives by adults. In fact, contests probably form a vital part of children's cultural development in Japan as they undoubtedly do elsewhere.

NOTES

1 An early example of the use of this word, according to the *Nihon Kokugo Daijiten*, was in the Tokyo newspaper the *Nichinichi Shinbun* in November 1881.
2 An interesting discussion appeared recently in Japanese about the introduction of the notion of equality and 'fair play' into Japanese sports and contests in the nineteenth century, largely from England (Shirahata 1989: 177–80).

3 For a detailed discussion of examples of these socialisation
practices, and how they are achieved see Hendry (1986).
4 I have discussed this point in Hendry (1987: Chapter 12),
where reference is made to several works on the subject,
among them Haley (1982); Henderson (1965); and Kawashima (1963, 1967). According to Haley, for example, the
rate of conviction in cases which go to trial is 99.9 per cent,
but in less than 3 per cent of those cases do the courts
impose a prison sentence, and in 87 per cent of those cases,
the terms are less than three years. Furthermore, two-
thirds of the jail sentences are regularly suspended, so that
less than 2 per cent of all those convicted of a crime ever get
imprisoned.
5 This is in fact only one of a variety of games played with the
fist in Japan, although possibly imported from China
around the seventeenth century (*Kodansha Encyclopaedia*,
IV). They are collectively known as *ken* (literally, fist), and
may also be played by adults, typically during drinking
parties.
6 The following information is drawn from a short book
about *sumo* published by the Japan Sumo Organisation
(*Nihon Sumo Kyōkai*), available to visitors to the Sumo
Museum in Tokyo, and from Cuyler (1985).
7 This area is the Yame region referred to earlier. I carried
out fieldwork there from 1975 to 1976, and again, more
briefly, in 1979, 1981, 1987, and 1988.
8 During certain periods, *sumo* was banned because it
became too rough and dangerous. The modern ritualised
version developed out of conscious efforts to make the
activity a socially acceptable spectable (Cuyler 1985: 59–
68).
9 An amusing series of short stories depicting this tension
among company employees is available in English
translation in Genji (1980).

REFERENCES

Cuyler, P. L. (1985). *Sumo: From Rite to Sport*. New York
and Tokyo: Weatherhill.
De Mente, Boye (1983). *The Whole Japan Book*. Phoenix,
Arizona: Phoenix Books.
Genji, Keita (1980). *The Lucky One and other Humorous
Stories*, trans. Hugh Cortazzi. Tokyo: Japan Times.
Griffis, W. E. (1874). The Games and Sports of Japanese
Children. *Transactions of the Asiatic Society of Japan* 2,
140–58.
Haley, John O. (1982). Unsheathing the Sword: Law without
Sanctions. *Journal of Japanese Studies* 8/2, 265–81.

Henderson, Dan Fenno (1965). *Conciliation and Japanese Law*. Seattle: University of Washington Press.

Hendry, Joy (1986). *Becoming Japanese*. Manchester: Manchester University Press.

—— (1987). *Understanding Japanese Society*. London: Routledge.

Iwado, Tamotsu (1936). *Children's Days in Japan*. Tokyo: Board of Tourist Industry, Japanese Government Railways.

Johnson, Thomas Wayne (1975). *Shonendan: Adolescent Peer Group Socialization in Rural Japan*. Taiwan: The Orient Cultural Service.

Kawashima, Takeyoshi (1963). Dispute Resolution in Contemporary Japan. In *Law in Japan: the Legal Order in a Changing Society*, ed. A. von Mehren, pp. 41–72. Cambridge, Mass.: Harvard University Press.

—— (1967). The Status of the Individual in the Notion of Law, Right, and Social Order in Japan. In *The Japanese Mind*, ed. Charles A. Moore, pp. 262–87. Honolulu: The University Press of Hawaii.

Kodansha Encyclopaedia of Japan (1983). Tokyo: Kodansha.

Nakada, Kohei (1970). *Nihon no jido yuji* (*Children's Pastimes of Japan*). Tokyo: Shakai Shisosha.

Nelson, Andrew N. (1962). *The Modern Reader's Japanese–English Character Dictionary*. Rutland, Vermont, and Tokyo: Tuttle.

Nihon Kokugo Daijiten (1975). Tokyo: Shogakukan.

Nihon Sumo Kyōkai (NSK: The Sumo Organisation of Japan) (n.d.). *Sumo*, Tokyo.

Sekai Daihakka Jiten (1967). Tokyo: Heibonsha.

Shirahata, Yozaburo (1989). Born to Play? Modern Sports and Body Image in the Japanese Mind (in Japanese, with English summary), *Nihon Kenkyw̄* (*Bulletin of the International Research Center*) 1, 175–88.

Yoshida, Mitsukuni, Ikko, Tanaka and Sesoko Tsune (1987). *Asobi: The Sensibilities at Play*. Tokyo: Cosmos.

ELIZABETH BAQUEDANO

The Mesoamerican Ballgame: Symbolic Aspects

I INTRODUCTION

The ballgame (called *ullamaliztli* in Nahuatl) was practised in Mesoamerica (see map) from the Preclassic period (*c*.1200 BC) to the Spanish Conquest in 1519. It was played by two sides with a solid rubber ball in specially made courts (*tlachtli*) (see Figure 1). Its purpose was mainly ritual but there was also a public sporting spectacle accompanied by gambling. It frequently led to sacrifice of the defeated. In this chapter we will first mention several interpretations put forward by modern scholars and then discuss the religious significance and symbolic aspects attributed to the Mesoamerican ballgame — mainly the promotion of agricultural fertility.

The sixteenth-century Spanish chroniclers described the ballgame in detail but because of its religious connotations, they ordered the destruction of the masonry courts, of which, according to Taladoire (1981: 28), there were more than 556 courts in Mesoamerica and neighbouring areas.

II THE MESOAMERICAN BALLGAME: BACKGROUND, HISTORY OF RESEARCH, AND IDEOLOGICAL SIGNIFICANCE

The religious and ritual connotations of the ballgame were recognised by the sixteenth-century chroniclers but they hardly understood its significance. Modern scholarship has to rely on the iconography of the ballgame in order to understand its meaning. That is, through the study of painted pottery, stelae, or sculpture in general — ballrings, ritual paraphernalia (*hachas, palmas*, etc.), and, obviously, painted manuscripts (Codices). Of great importance is also the *Popol Vuh* (Tedlock 1985), the sacred book of the Quiche Maya, written down after the conquest.

The German scholar, Eduard Seler in his commentaries on the Borgia Codex group was the first to come up with an interpretation

PACIFIC OCEAN

Shaded areas: altitudes over 600 m

```
|———|———|
0    75   150
     miles

|———|———|
0   100  200
      km
```

Map of Mesoamerica (After Porter Weaver 1981: 11).

BAQUEDANO

Figure 1 Schematic representation of the sacred precint of Tenochtitian:

of the significance of the ballgame. He suggested solar/lunar/astral interpretations, and concluded that the game's violent competition symbolised the battle between the forces of darkness and those of light, and was a re-enactment of the cosmic drama of the death and rebirth of the sun and other heavenly bodies such as Venus (Seler 1902–23, III: 308–10, IV: 15–6). He also suggested that the court could be interpreted as a cosmic metaphor for the earth and the four cardinal directions.

In 1948, Walter Krickeberg, agreeing with Seler about the diurnal and terrestrial meanings of the ballgame, further suggested a nocturnal celestial significance for the ballcourt. For Krickeberg, the ballgame expressed the conflict between the dual forces, light versus darkness. Lothar Knauth, in 1961, argued that the ballgame was a sacrificial cult related to the moon goddess and that its objective was the maintenance of agricultural fertility. He interpreted the frequent representations of decapitation rituals in ballgame iconography as the replication of the cosmic drama, in which the moon, personified by female deities (who particularly expressed the concept of agricultural fertility and whose ritual representatives were often decapitated), had to be immolated, that is, made to disappear, in order that the sun might rise.

In 1972, Esther Pasztory advanced the hypothesis that the ballgame symbolised the death and rebirth of the sun. The path of the ball represented metaphorically the daily and yearly trajectories of the sun; the ballcourt was the underworld, and also the night sky traversed by the sun between sunset and sunrise (Pasztory 1983: 124).

Marvin Cohodas (1975) hypothesised (on the basis of symbolic imagery in Mesoamerican art e.g. Maya) that the ballgame was performed on the equinoxes to influence, by sympathetic magic, the sun's descent into and subsequent ascent, reborn from the underworld. Cohodas (1978: 245) based his interpretation 'on the yearly cycle of the sun which descends into the underworld on the

---

Above the main pyramid with its two shrines there is a representation of Huitzilopochtli (the tribal god of the Aztecs) holding a fire-serpent on his right hand. On the left shrine there is the date 5 Lizard. On the one on the right the date 5 House. On both sides of the shrines there are standard-bearers. Below the main pyramid there is an image of Quetzalcoatl and a skull-rack underneath. Further down, the representation of the 'Divine Ball-court' (*teotlachco*). On both sides of the ball-court there are several buildings among them the School for the children of nobles *Calmecac* (left) and on the right the gladiatorial stone (*temalacatl*)). From: *Códice Matritense del Real Palacio*, fol. 269 r. after Sahagun (from Léon-Portilla 1978: 43)).

vernal equinox and ascends into the heavens on the autumnal equinox'. Cohodas, Baudez, and other scholars in the last twelve years have further demonstrated that the Mesoamerican ballgame myths, rituals, and inconography often have both strong astral/solar and agricultural fertility connotations.

Following Furst's (1982) interpretation of skeletal symbolism among the Mixtecs, the present writer (1988, 1989) has demonstrated that the skull symbolism is related to many concepts and ideas, among them agricultural symbolism (death and regeneration) and that this applies directly to the symbolism and iconography of the ballgame, both in sculptural reliefs associated with the ballgame and in Codices. For example, in Codex Magliabechiano (80 recto) there is a representation of a ballcourt (*tlachco*) in the form of a capital I, depicting two players, one at each end of the court. Three skulls in profile form the central line, and there are two more skulls near each of the players. The stone-rings (*tlachtemalacatl*) are located in the middle of the court. The one on the left is painted red and the other an ochre colour. The player on the left holds the ball. The skull is associated with agriculture not only among the Mixtecs but also among the Maya. It is a pan-Mesoamerican symbol linked with the earth, containing regenerative forces that connected it to agricultural production and fertility in general. This would make sense in that one of the most important objectives of the ballgame was the promotion of agricultural fertility through the notion of death and regeneration.

III THE BALLGAME IN THE MAYA AREA

In the lowland Maya region, important discoveries involve historical rulers (Figure 2). These include ballgame scenes on Late Classic ceramic vessels, relief sculptures that represent players in action, and references to the game in hieroglyphic texts. In the early 1970s, Michael Coe observed an elaborate and highly patterned symbolism used by Maya artists working on pictorial pottery (Coe 1973: 16). He interpreted much of the iconography (including ballgame scenes on Late Classic pots) in terms of the adventures of the Hero Twins of the *Popol Vuh*. The story goes that two men, called 1 Hunahpu and 7 Hunahpu (the first fathers), were the best players on earth. They practised tirelessly, but the noise of the bouncing rubber ball disturbed the lords who lived in the Maya underworld (*xibalba*). The lords of the underworld summoned the brothers to a ballgame there. The gods won the game by deception and the two brothers were sacrificed. The body of 7 Hunahpu was buried in the ballcourt and they hung 1 Hunahpu's head in a calabash tree as a symbol of their victory.

Figure 2 Carved Maya panel showing two richly costumed ballplayers tossing a glyph ball under glyph block indicating that one of the players is a ruler. Ball player on left wears hip and chest protector with hacha attached and arm and knee padding: the reclining player wears a chest protector and arm and knee padding. (After Schele and Miller 1986: 257).

One day the daughter of an underworld lord walked past the calabash tree and spoke to the skull, whereupon it spat into her hand, miraculously making her pregnant. Her outraged father ordered that she be sacrificed, so she fled to the middleworld where she sought refuge in the house of 1 Hunahpu and 7 Hunahpu, where the twins' mother lived. She granted her permission to stay but she tested the divinity of the young woman. The latter proved herself by harvesting a full net of corn from a single plant. She then gave birth to twins named Hunahpu and Xbalanque. These two discovered the ballgame and produced the bouncing noise of the rubber ball — upsetting the gods who lived below in the underworld, so that once again the gods invited the twins to compete against them. The account concludes with the defeat of the gods of the underworld, and the apotheosis of Hunahpu and Xbalanque, taking their places as the heavenly bodies identified as the sun and the moon. The myth has been interpreted as the re-enactment of the combat between the forces of darkness and the forces of light. Pasztory (1972: 445) has interpreted the myth as a reference to the annual and diurnal solar cycles. Mary Miller and Stephen Houston (1987) stress the importance of the human sacrificial theme in which the contest often culminated.

The Maya version of the ballgame at the time of the Conquest is described in the *Popol Vuh*, and its role in the mythology of death and sacrifice is demonstrated. More a mythico-religious ritual than a sport, the game usually culminated in human sacrifice, normally decapitation. Thus it can be treated as a formal means of selecting the best qualified 'victim', i.e. the loser. The cult may have served purposes of divination and the prognostication of extraterrestrial events, but through time it was increasingly manipulated for worldly political purposes as well. Esoteric meanings attached to the movements of the rubber ball, the layout of the courts, the action of the players, the outcome, and the sacrifices after the game fundamentally pertained to concepts of maintaining primary cosmic cycles — particularly the vernal and autumnal equinoxes — and seasonal agricultural fertility. The pervading dualities of dry season/rainy season, sky/underworld, day/night, sun/moon, morning/evening (Venus), and most especially death/rebirth, were emphasised. There was a preoccupation here with the underworld including the passage and transformation of sacrificed ballplayers, which primarily symbolised the diurnal death and rebirth of the sun (and the moon). The sacrifice of the sun (in the west) in the guise of a 'privileged' ballplayer helped to assure its successful underworld passage and ultimate transformation and rebirth (in

Figure 3 Panel from Aparicio, Veracruz. Portrays a seated ballplayer complete with *palma* and hammerstone or stone weight. His head has been severed and seven intertwined serpents sprout from his neck. (After Macazaga Ordoño 1985: 61).

the east). This was metaphorically and crucially equivalent to the regeneration of maize and vegetation and life itself.

IV THE GULF COAST AREA

The Late Classic site of El Tajin, Veracruz, in the Gulf Coast area, was of pivotal importance for the practice of the ballgame. At this location seven ballcourts have been found so far. The north and south courts have six low-relief carved panels on the side walls elaborately portraying mythology. Wilkerson (1987) has analysed the six elaborate scenes at the South Court. They depict stages of ballgame practice and myth, from pre-game preparations and post-game decapitations in their corner panels, to the underworld aftermath in the central panels. The reliefs show players wearing items of the ballgame such as yokes and palmas. There are likewise bench profiles and balls, various attendants, and patron deities. Among the last, Venus, rain, and pulque gods are predominant. These in turn carry strong cosmic and fertility connotations. One stela (Figure 3) also from Veracruz (Aparicio) depicts a decapitated ballgame player. From his neck issue seven intertwined serpents. The symbolism is with earth, blood, and agricultural fertility in general. The same type of depiction appears at distant Chichen Itza (Yucatan). The six bench panels at the Great Ballcourt there each depict processions of fully accoutered ballplayers wearing yokes and palmas and carrying handstones, seven members to a team. Directly facing the ball in the centre (containing a skull motif) is a half-kneeling decapitated player with six serpents and one flowering vine emerging from the neck, while the captain of the opposing team, on the other side of the ball, holds the knife and the trophy head.

The complex of yoke (or *yugo*), *hacha*, and *palma* originates in the Veracruz region and diffuses southward in Classic times. Among the human burials associated with stone yokes, several known interments appropriately include the actual sacrificial skulls. Leyenaar and Parsons (1988: 49) mention several examples of this, among them Santa Luisa near El Tajin, and at Cerro de las Mesas and in Omealca. This suggests that the stone equipment was often buried along with dead players, perhaps to allow them to continue their cosmic contest in the underworld (cf. the *Popul Vuh* mythology).

V POTTERY

Pottery figurines are particularly prominent in the Gulf Coast area where they graphically portray all elements of ball players regalia. They are also plentiful in the Maya area, some of them still bearing colour remains. Pottery motifs include images of Tlaloc (the god of rain) and of players grasping knives and trophy heads. Some examples feature a decapitated player with six serpents plus a central device spurting from its neck, just as seen on the Veracruz 'Aparicio' panels and on the bench panels at Chichen Itza. Some examples from the Maya region show richly attired players in action or holding the rubber ball.

VI THE GULF COAST AREA PANELS

In the Gulf Coast area of Veracruz, especially at the site of Vega de Alatorre in Veracruz, there are relief panels showing decapitated ballplayers. An example from the Rijksmuseum voor Volken-kunde, Leiden (Leyenaar and Parsons, 1988: 205) shows a decapitated ballplayer with a yoke, two palmas, a knee pad, and a *manopla* (gauntlet) on his right hand. Seven interwined serpents sprout from his neck. The fact that there are seven serpents depicted is meaningful. The Aztec goddess of maize and sustenance was Chicomecoatl (the name meaning Chicome = seven and coatl = serpent) which is exactly what is depicted in the panel under discussion. Although the panel is of much earlier date (AD 400–700) than the Aztecs, it is likely at least that the Aztecs borrowed the concept of serpents and agricultural fertility from their predecessors — but they certainly followed the practice even more. For example, the corpus of Aztec sculpture includes a rattlesnake tail with maize cobs which seem to grow from the serpent's tail (Baquedano 1989). Agriculture and snake symbolism go hand in hand. After all, one of the aspects of the ballgame was linked with agriculture. In Mesoamerica the serpent itself was considered the most important fertility symbol.

Decapitation generated streams of blood that fell on the earth. So the connection with fertility becomes clear. Beheading was an ancient practice in Mesoamerica as confirmed, for example at Tikal (Guatemala), by the archaeological record (Coe 1965: 19, 21).

VII THE BALLGAME: AGRICULTURAL FERTILITY AND OTHER SYMBOLIC ASPECTS

The main evidence for the relationship with agriculture comes from ballgame contexts. Cylindrical tripod vases made in Teoti-huacan style and found at the site of Escuintla in Guatemala depict ballplayers holding decapitated human heads. The victim is shown with serpents sprouting from his neck (Hellmuth 1975: 16–29). A direct association between the ballgame and decapitation is clearly shown in these vases as well as in reliefs from Aparicio, Veracruz in the Gulf Coast.

On the great ballcourt panels at Chichen Itza (Prem and Dyckerhoff 1987: 98–9), instead of a seventh, central serpent there is a magnificent plant, clear proof of the assimilation of serpents with vegetation. In Mesoamerica the snake was associated with fertility and agriculture, as has already been mentioned.

The Mixtec Codex Nuttall (folio 3) shows a decapitated woman with a mask of Tlaloc standing on a ballcourt; above her, there is a depiction of an opossum. The opossum was associated with the ballgame, fertility, and decapitation. Because the female opossum gives birth to several creatures, she was associated with abundance and fertility. Again, the relationship with agriculture (Tlaloc), fertility (the opossum), and the ballgame is clear.

Another link between the ballgame and agriculture comes from the Toltecs. It is said that before the fall of the Toltec capital of Tula $c.987$ AD, Huemac the last ruler of the Toltecs was visited by some Tlaloqueh (rain gods) to invite him to play a ballgame. It was agreed that the winner would obtain precious quetzal feathers and jades. The winner was Huemac. He expected to obtain quetzal feathers and jades; instead he was given green maize cobs and green maize leaves. Huemac angered by this said that that was not what had been agreed, and refused to accept the maize cobs and leaves. The myth has it that as a result of his arrogance the gods of rain withdrew what they considered to be their precious jades and quetzal feathers (i.e. maize). After this event followed a period of four years of famine which led to the final fall of Tula (Códice Chimalpopoca 1975). The myth clearly illustrates the connection between the ballgame and agriculture, and the punishment to those who reject the precious quetzal feathers and the jades of the gods of rain, maize.

VIII THE BALLGAME AMONG THE AZTECS

The Aztec ballgame was played with various intentions. Other than the religious aspect, it was played in everyday life for gambling. It was an excuse for huge bets. Large quantities of clothes, gold, and even slaves, changed hands. The game was played by the ruling class and members of the elite. Many of them ended their days in ruin. The game was played by two men or teams consisting of up to four players. Players were not allowed to touch the ball with either hands or feet, but only with the lower legs or upper arms, knees and hips. To protect themselves from the ball, players wore protective garments on their arms, knees, and legs. The aim of the game was to propel the ball through the stone rings tenoned in the side walls. Scoring was difficult and, according to sixteenth-century Chroniclers, many players died of exhaustion before scoring any points. Players could score points by making contact with the rings or by directing the ball into a goal area of the court (Schele and Miller 1986: 243). The losers were sacrificed.

Among the Aztecs, the ballgame was also related to beheading and fertility (Krickeberg 1966). An Aztec ballcourt ring at the National Museum of Anthropology (no. 11–3512) depicts, in relief, a man holding a severed head.

In Codex Borbonicus (folio 19) there is a representation of Xochiquetzal (goddess of flowers and agriculture) in front of a ballcourt and a decapitated man. Xochiquetzal is a youthful aspect of the Earth goddess. In the centre of the ballcourt is a skull with a flow of water. This seems to be a direct allusion to the Aztec and probably Mesoamerican belief in the ballgame as an instrument for ensuring the alternation of the dry and rainy seasons. During the wandering period (*c.* AD 111–1325) of the Aztecs they stopped at Coatepec, where they built a temple to their patron god, Huitzilopochtli, and a ballcourt with a hole in the middle in which they put water; it was thanks to the water issuing from the hole that trees and vegetation, fish and game appeared abundantly. But when Coyolxauhqui and the 400 Huitznahua (Southerners) wished to stay there, instead of continuing the journey to the promised land, Huitzilopochtli killed them during the night; and he cut Coyolxauhqui's head and extracted her heart exactly over the ballcourt's hole. As a result, the following morning the water disappeared and the animals too (Tezozomoc 1944: 228–9). The dry season, equated with day, succeeded the rainy season, equated with night. In this myth, the association between ballgame and fertility is very clear, but Coyolxauhqui's death, though also linked with agriculture, appears to have a negative effect on it. The explanation may be that Coyolxauhqui represents in this myth, as demonstrated by Seler (1902–23, III: 238–9) the old waning moon. In Aztec ritual, her death was re-enacted during the month Titl when a slave impersonating the 'Old Lady' was sacrificed and beheaded in order to have the dry season succeed the rainy one; exactly half a year later, in *Huey Tecuilhuitl*, a young goddess, Xilonen, was decapitated with completely different purpose — this time to have the rainy season succeed the dry season. The old moon and the young one were opposed to each other like sterility to fecundity (Graulich in press).

The alternation of the seasons, rainy and dry, is referred to in a version of the story of Huitzilopochtli at Coatepec, as part of the

Mexica migration narrative. The Mexica sojourn at Coatepec in these accounts began with Huitzilopochtli creating a lush paradise by damming a river or creating a source of water. Thus Huitzilopochtli, a deity with solar aspects, is shown here to be a master of terrestrial waters and is thereby opposed to Tlaloc, with whom he shared the summit of the Great Temple of Tenochtitlan. Tlaloc was the master of celestial waters (rain) but his name refers to the earth (*tlalli* = made of earth), and he was said to live under the earth — a further opposition between him and the sun of the daytime sky.

When Huitzilopochtli decided to punish those among the Mexica who wished to stay at Coatepec rather than continue their journey, he killed the traitors in the middle of the night, opened their chests, and ate their hearts. He cut off Coyolxauhqui's head and threw it into the centre of the ballcourt, at a place known as the *itzompan*, the place of the skull. Then he broke the dam, letting all the water run out to create a desert devoid of plant and animal life. Durán (1964: 21) describes the event as follows:

> At midnight, when everything was quiet, the people heard a great noise in a place called Teotlachtli, the Divine Ball Court, and in the Tzompanco, Skull Rack, dedicated to the deity [Huitzilopochtli]. When the morning came they found the principal instigators of the rebellion dead there, together with the woman called Coyolxauhqui. All their chests had been opened and the hearts removed.
>
> From this came the accursed belief that Huitzilopochtli ate only hearts, and thus was established the practice of sacrificing men . . ..

The Teotlachco of Tenochtitlan was the site of the annual dramatic, ritualist re-enactments of this myth. During the fifteenth Aztec 'month', Panquetzaliztli, there was a feast, dances, songs ceremonies, and food offerings dedicated to the Mexican tutelary deity, Huitzilopochtli. A number of war prisoners were sacrificed to the god (Sahagun 1950–78, II: 27). Before more general human sacrifices took place, however, a special introductory ceremony was conducted in the great ballcourt.

The Great Temple of Tenochtitlan, with its twin shrines, represents a juxtaposition and complementarity of elements and seasons: Huitzilopochtli, the celestial solar god/the dry season, and Tlaloc, the rainy season associated with underworld and terrestrial god of agricultural fertility. The alternation of the seasons and the equilibrium between the two seemed to be two principles directly related not only to night and day but also to the agricultural concerns that were important aspects inherent in the ballgame. It is my contention that one central concern of the

Mesoamerican ballgame was agricultural fertility; but in the case of the Aztecs there was also a political moral. Those who opposed the wishes of Huitzilopochtli, exemplified by the myth of Huitzilopochtli's birth, die: they were deprived of water and food, slain and abandoned. Sahagun's version of the story emphasises the beginning of the daily cycle of the sun rising out of the earth, initiating the alternation of day and night. Thus the head of Coyolxauhqui remained on top of the mountain close to the sky, where the moon should be, as an astral body of the night sky. In the latter accounts, on the contrary, the head of Coyolxauhqui was placed in the ballcourt, which is an entrance of the underworld (Gillespie in press), the counterpart of the *coatepetl* ('place of the hill of serpents') the entrance to the realm of the day sky. More precisely, her head was placed in a hole containing water in the ballcourt (Tezozomoc 1944: 229). As the moon, Coyolxauhqui is related with water (Hunt 1977: 139). The result here was a permanent conjunction of earth and sky (represented by the moon) (Gillespie in press).

IX TEOTLACHCO, THE DIVINE BALLCOURT OF TENOCHTITLAN

At the time of the Conquest, one of the functioning ballcourts was the so-called Teotlachco (place of the 'Ballcourt of the Gods' or 'Divine Ballcourt'). The importance of this structure is indicated by the fact that it is one of the buildings illustrated by Sahagun (1984, II: plate 16) on his plan of the Great Temple ceremonial precinct. According to this plan, the Teotlachco lay opposite and in front of the Great Temple of Huitzilopochtli and Tlaloc. Shown between the Teotlachco and the Great Temple are the Temple of Quetzalcoatl and the *tzompantli* (skull-rack) a wooden structure for the display of human skulls (León-Portilla 1978: 43).

Hernando Alvarado Tezozomoc's (1944: 11–4) account of the legendary peregrination of the Aztecs from Chicomoztoc to Tenochtitlan throws light on the symbolic meanings of the Teotlachco. During their travels, the Aztecs stop at the place known as Coatepec near Tollan (Tula). At Coatepec was located a ballcourt called Teotlachco, which had a kind of well or spring at its centre, known as *Itzompan* ('place of skulls').

A version of the ballgame (*Ulama*) has survived in Mexico in the modern state of Sinaloa with purely sporting connotations (see Leyenaar 1978).

X CONCLUSIONS

The movement of the ball symbolised the heavenly bodies of the sun and moon or Venus as well as the alternation of light and darkness. According to the art and inscriptions, death and human sacrifice were frequently the outcome of the Maya and Aztec ballgame. In some cases, death is shown explicitly; in others it is implied. In Post-Classic images, such as the ballgame scenes at Chichen Itza, defeated ballplayers are decapitated.

The *Popol Vuh* is of further interest as it reveals that skulls are sources of fertility and that they have as much to do with life-giving properties as with death. The *Popol Vuh* also testifies that losers were beheaded and that skulls were used as symbols of victory. Among the Aztecs, the idea of life and regeneration was present in many contexts, especially in agriculture. The skull symbolised victory as well as life-sustaining qualities.

In short, although the game underwent many changes, the ideas of the movement of the stars and of agriculture persisted throughout Mesoamerica. This is attested in both the archaeological and the written records.

ACKNOWLEDGEMENT

I am most grateful to Emily Lyle for her original invitation to give a paper at the Contests conference and to Warwick Bray and Nicholas James for reading this paper and for their editorial comments.

REFERENCES

Baquedano, Elizabeth (1988). Iconographic Symbols in Aztec Elite Sculptures. In *Recent Studies in Pre-Columbian Archaeology*, eds N. Saunders and O. de Montmollin, pp. 191–204. Oxford: BAR International Series 421.
—— (1989). Aztec Death Sculpture. Unpublished Ph.D. dissertation, University of London.
Baudez, Claude (1984). Le roi, la belle et la maiz: Images du jeu de balle Maya. *Journal of la Societé des Americanistes* LXX, 139–52.
Codex Borbonicus (1981). Commentary E. T. Hamy. Mexico: Siglo XXI.
Codex Magliabechiano (1983). Ed. Elizabeth H. Boone. Berkeley: University of California Press.
Codex Nuttall. (1975). Ed. Zelia Nuttall, with text by Arthur G. Miller. New York: Dover Publications Inc.
Códice Chimalpopoca (1975). Ed. Primo Feliciano Velázquez. México: Instituto de Investigaciones Históricas, Universidad National Autónoma de México.
Coe, Michael (1973). *The Maya Scribe and his World*. New York: Grolier Club.
Coe, William (1965). Tikal: Ten years of Study of a Maya Ruin in the Lowlands of Guatemala. *Expedition* 8/1,

5–56. Philadelphia: University Museum, University of Pennsylvania.

Cohodas, Marvin (1975). The Symbolism and Ritual Function of the Middle Classic Ball Game in Mesoamerica. *American Indian Quarterly* 2/2, 99–130.

—— (1978). *The Great Ball Court at Chichen Itza, Yucatan, Mexico.* New York: Garland Publishing Co.

Durán, Diego (1964). *The Aztecs: The History of the Indies of New Spain,* trans. Doris Heyden and Fernando Horcasitas. New York: Orion Press.

—— (1971). *Book of the Code and Rites and the Ancient Calendar,* eds and trans. Fernando Horcassitas and Doris Heyden. Norman: University of Oklahoma Press.

Furst, Jill (1982). Skeletonization in Mixtec Art. In *A Re-Evaluation in the Art and Iconography of Late Post-Classic Central Mexico,* ed. Elizabeth H. Boone, pp. 207–25. Washington, DC: Dumbarton Oaks, Trustees for Harvard University.

Gillespie, Susán D. (in press). Ballgames and Boundaries. In *The Mesoamerican Ballgame,* eds Vernon Scarborough and David Wilcox. Tucson: University of Arizona Press.

Graulich, Michel (in press). Les grandes statues azteques dites de Coatlicue and Yollotlicue. In *Homage à Pierre Duviols.*

Hellmuth, Nicholas (1975). The Escuintla Hoards: Teotihuacan Art in Guatemala. *F.L.A.A.R. Progress Reports* (Guatemala City) 1/2, 5–19.

Hunt, Eva (1977). *The Transformation of the Hummingbird: Cultural Roots of a Zinacantecan Mythical Poem.* Ithaca: Cornell University Press.

Kampen, Michael (1972). *The Sculptures of El Tajin Veracruz, Mexico.* Gainesville: University of Florida Press.

Knauth, Lothar (1961). El Juego de Pelota y el Rito de la Decapitación. *Estudios de Cultura Maya* 1, 183–98.

Kowalski, Jeff (n.d.). Astral Deities and Agricultural Fertility: Fundamental Themes in Mesoamerican Ballgame Symbolism at Copan, Chichen Itza, and Tenochtitlan. Paper presented at the 46th International Congress of Americanists, 4–8 July 1988, Amsterdam.

Krickeberg, Walter (1948). Das mittelamerikanische Ballspiel und seine religiose Symbolik. [Spanish translation 1966: El Juego de Pelota Mesoamericano y su Simbolismo Religioso. *Traducciones Mesoamericanistas* 1, 191–313. México: Sociedad Mexicana de Antropología.]

León-Portilla, Miguel (1978). *México-Tenochtitlan: su espacio y tiempo sagrados.* México: INAH, 477–502.

—— (1988). El Maíz nuestro sustento, su realidad divina y humana en Mesoamérica. *América Indígena* XLVIII/3.

Leyenaar, Ted (1978). *Ulama: The perpetuation in Mexico of the Pre-Spanish Ball Game Ullamaliztli*. Leiden: E. J. Brill.

Leyenaar, Ted and Leé Parsons (1988). *Ulama: The Ballgame of the Mayas and Aztecs 2000 BC–AD 2000*. Leiden: Spruyt, Van Mantgem.

Mecazaga Ordaño, César (1985 [2nd. edition]). *El Juego de Pelota*. México: Editorial Innovación.

Miller, Mary and Stephen Houston (1987). The Classic Maya Ballgame and its Architectural Setting. *RES* 14, 45–65.

Moser, Christopher L. (1973). *Human Decapitation in Ancient Mesoamerica*. Studies in Pre-Columbian Art and Archaeology XI. Washington, DC: Dumbarton Oaks, Trustees for Harvard University.

Pasztory, Esther (1972). The Historical and Religious Significance of the Middle Classic Ball Game. *Sociedad Mexicana de Antropología, XII Mesa Redonda*, 441–55.

—— (1983). *Aztec Art*. New York: Harry N. Abrams.

Porter Weaver, Muriel (1981 [2nd. edition]). *The Aztecs, Maya and Their Predecessors: Archaeology of Meso-america*. New York: Academic Press.

Prem, Hans J. and Ursula Dyckerhoff. (1987). *Le Méxique ancien. L'Histoire et la culture des peuples de la Mésoamérique*, prés. Michel Graulich. s.l.: Bordas/ Civilisations.

Robicsek, Francis (1981). *The Maya Book of the Dead: The Ceramic Codex: the Corpus of Codex Style Ceramics of the Late Classic Period*. Charlottesville: University of Virginia Art Museum.

Sahagún, Bernardino de (1950–78). *Florentine Codex. General History of the Things of New Spain*, trans. Arthur J. O. Anderson and Charles E. Dibble, 13 parts. Santa Fe: School of American Research and the University of Utah.

—— (1984). *Códices Matritenses de la Historia General de las Cosas de la Nueva Espana*. (Ed. Manuel Ballesteros-Gaibrois), 2 vols. Colección Chimalistac de libros y documentos acerca de la Nueva España 19. Madrid: Ediciones José Porrúa Turanzas.

Schele, Linda and Mary Miller (1986). *The Blood of Kings: Dynasty and Ritual in Maya Art*. Fort Worth: The Kimbell Museum.

Seler, Eduard (1902–23). *Gesammelte Abhandlungen zur Amerikanischen Sprach und Alterhumskunde*, 5 vols. Berlin: A. Asher & Co. and Behrend & Co (Reprint 1980–81: Akademische Druck- und Verlagsanstalt, Graz.).

Taladoire, Eric (1981). *Les Terrains de jeu de balle (Méso-amérique et Sud-ouest des Etats-Unis)*. México: Estudios

Mesoamericanos, Serie II: 4, Misión Arqueológica y Etnológica Francesa en México.

Tedlock, Dennis (1985). *Popol Vuh: The Definitive Edition of the Mayan Book of the Dawn of Life and the Glories of the Gods and Kings*. New York: Simon and Schuster.

Tezozomoc, Hernando Alvarado (1944). *Crónica Mexicana escrita hacia el año 1598,* ed. Manuel Orozco y Berra. México: Editorial Leyenda.

Wilkerson, Jeffrey (1987). *El Tajín: Una Guía Para Visitantes.* México: Universidad Veracruzana.

ALICE B. KEHOE

Contests of Power in Blackfoot Life and Mythology

I BACKGROUND

The Blackfoot are a confederacy of bands occupying the North-western American Plains, now southern Alberta, Canada, and adjacent north-central Montana. At the time of first European contact, 1690, they lived as nomadic bison hunters in southwestern Saskatchewan. Archaeological evidence indicates that they probably used this territory at least as early as the fifteenth century, and possibly many centuries before. As speakers of a Central Algonkian language, the Blackfoot are the western frontier of this language group that includes the Blackfoot neighbours the Gros Ventres (Atsina) and the Cree.

Subsistence for the pre-reservation Blackfoot was based on communal hunting of bison through impounding herds. Cervids, small game, birds, roots and tubers, and berries supplemented bison meat, and both hides and bones of slaughtered animals were transformed into a variety of objects including the tipi cover, clothing, containers, and implements. When horses were obtained through trade in the mid-eighteenth century, they were used primarily as pack and riding animals in travelling; although bison were then pursued on horseback, communal pounds continued in use until the extermination of the wild bison herds in the 1870s.

Blackfoot social organisation centred on small bands of several families, often polygynous in the eighteenth and nineteenth centuries when endemic warfare caused heavy mortality among the men. Each band recognised male and female leaders, frequently older persons who had demonstrated superior skills in managing economic and social resources. Anyone who was dissatisfied with a band's management could leave and join another band, and even multi-ethnic camps were not uncommon on the Northwestern Plains. At the height of summer grazing, bands gathered for trading, games, ceremonies, matchmaking, adjudication of disputes, and political alliances. During the remainder of the year, bands lived separately, near a series of pounds. Young men in particular sought adventure, in the historic period, in raiding for horses as well as in extended travels to distant regions.

II POWER

Blackfoot believe the universe is animated by an Almighty Power (Kitche Manitou) manifested through phenomena perceptible to humans. The Sun and Thunder are the most impressive manifestations of Power; anomalous phenomena such as water mammals (beavers and otter) or rocks that seem to move are other strong manifestations of Power, and so were the physically powerful bison and bears, and the rattlesnake, whose bite could quickly cause death. None of these manifestations is worshipped in itself, but is invoked in prayer as a link to the Almighty Power. Many less impressive phenomena figure in rituals because they were seen, in visions or in actuality, signalling Power by behaving anomalously. Singing prayers over symbols of Power invokes it.

In Blackfoot eyes, humans are among the less powerful beings. Only the beneficience of the Almighty gives humans life and the wherewithal to sustain it. From time to time, Power granted gifts to worthy humans, who passed on the gift in the form of medicine objects or bundles. The most important medicine bundles symbolise Power manifested through Sun and Thunder, and the most common medicine object is probably the *iniskim*, a small object, usually a fossil ammonite shell (found in Blackfoot territory), resembling a bison and believed to bestow prosperity and health upon the family that honours it. Knowledge of the *iniskim* and the bison-calling ritual that accompanied it is said to have been given originally to a woman searching for food for her family. This and other food-getting rituals are understood to be means of 'crying' before the Almighty and Its manifestations the food organisms, so that the Power 'pities' (Blackfoot speakers' translations) and is generous towards the human petitioner. Thus hunting is *not* ordinarily conceptualised as a contest between man and beast, but as benevolence.

Contests of power do occur. They are in essence contests between beneficiaries of Power. Although the Almighty favours generosity reflecting Its benevolence, beings enjoy free will and some may exercise power towards selfish ends. Both myths and actual incidents describe contests of power between a protagonist and one who uses power for evil. 'Evil' is, of course, a label coloured by the storyteller's loyalties and opinions. The Blackfoot cultural tradition does not favour dichotomies or oppositional dualism, so both mythical characters (who are understood to have been historical persons) and real individuals are seen to be potentially more or less inclined towards good, or evil, behaviour.

Sorcery can be practised by persons with medicine power, i.e. adepts who have gained uncommon power through vision bestowal, either directly in their own vision experience or, more likely in Blackfoot society, by purchasing medicine objects and the accompanying ritual. An example of sorcery is this incident recorded in 1939 on the North Piegan reserve in Alberta by Ruth and Oscar Lewis:[1] It happened, about 1937, that a man had traded a leather rope for a horse. The former owner of the horse then was offered cash for the animal, and asked for its return. The new owner refused, and the former owner walked twice around him and prophesied he would suffer bad luck. The next month, the new owner's daughter died. His father's sister's husband, a member of the Horn Society (a shamans' secret society), invoked Horn power and prayed to it to punish his wife's nephew's persecutor. To activate the curse, the Horn man placed a bone behind the persecutor's house. After that, the persecutor became poor and friendless — according to the Horn man, who told the story (Lewis and Lewis 1939: 20 August).

A contest of power is described in another incident from the Lewises' fieldnotes. This event is said to have occurred around 1870, shortly before the treaty with Canada establishing the Blackfoot reserves in Alberta. In this incident, a man named Big Snake organised a raiding party against the customary enemies, the Crows. Big Snake dreamed of a pure white horse, and instructed his followers that if any obtained such a horse, he should present it to the leader. One of the party, Seeing Far, did capture a white horse, but decided to keep it. Big Snake cursed Seeing Far, prophesying that he would never steal another horse. On a subsequent futile raiding journey, Seeing Far befriended two crows (birds) and was rewarded by them telling him to seek the assistance of a Holy Woman (woman who 'made' a Sun Dance) to lift the curse. Seeing Far's comrade, Rain Chief, was brother to the Holy Woman vouchsafed in the vision given by the crows. The men returned home and Seeing Far presented a horse packed with gifts to the Holy Woman, asking her to help him. She had him cleanse himself in the sweatlodge, take some feathers as tokens, and advised him to present a bay horse to Big Snake following his first successful raid, which soon took place. The Lewises' informant commented, 'A woman could have more power than a medicine man' (Lewis and Lewis 1939: 20 August).

III THE STORY OF KUTOYIS

The most extended Blackfoot exposition of the theme of contests
of power is the myth of Kutoyis ('Blood-Clot', referring to
menstrual or placental blood[2]) (for versions and related data see
Black Boy 1973; Grinnell 1892; Thompson 1957 [Motif T541.1.1,
v5:396]; Uhlenbeck 1911; Wissler 1911, 1912, 1936; Wissler and
Duvall 1908). Here is Uhlenbeck's (1911) transcription, condensed:

A man with three sisters as his wives camped with his parents-
in-law. Contrary to proper behaviour, the man abused his
parents-in-law, denying them food and requiring his father-in-
law to assist him by driving bison into their pound while the
son-in-law waited to shoot them. Only the youngest daughter
dared defy her unnatural husband and secretly give meat to
her parents.

One morning as the old man followed the bison toward the
pound, he saw a large clot of blood on the trail, and quickly
put it into his quiver, hiding it from his cruel son-in-law.
When he returned to his tipi, the old man gave the bloodclot
to his wife, who put it into the pot over the fire. Something
suddenly cried from the water: a child. The old woman pulled
him out and wrapped him up. When the son-in-law heard a
baby crying from his in-laws' tipi, he sent his wives one at a
time to discover whether the child was a girl whom he might
marry. The old people said it was, and the son-in-law sent
over some bones that the baby might have soup.

That night, the baby told his mother, the old woman, 'Hold
me to all these lodge poles. Begin by holding me by the
eastern one by the door.' After they had circled the lodge, the
child had become a young man, and asked for food. The old
couple explained to him the cruelty of his brother-in-law.
Kutoyis told his father that before dawn, they would go drive
bison into the pound.

When the old man drove in the animals, Kutoyis killed the
fattest cow, and telling his father to rest, the young man began
butchering it for him. The son-in-law arrived, enraged
because the old man had not waited for him. Kutoyis told the
old man, 'What your son-in-law says, you must repeat right
after him.' The son-in-law angrily said, 'Aha, there is nobody
to prevent me from killing you!' This the old man repeated,
'Aha, there is nobody to prevent me from killing you!' and
continued eating the choice meat. The son-in-law, 'That one
eating is living the last of his life!' The old man, to Kutoyis
who was hidden, 'Jump up, he is very close.' Kutoyis: 'Wait,
talk back to him.' The old man, to his son-in-law, 'He is living
the last of his life, who is coming this way.' The son-in-law

shot at the old man, but missed, Kutoyis jumped up and killed the wicked son-in-law.

Returning to camp, Kutoyis asked his parents which of their daughters pitied them. They answered, 'The youngest one.' Kutoyis then killed the two other women and instructed his youngest sister to care for their parents. 'Tomorrow', he said, 'I shall go higher up the river, visiting.'

When he reached the next camp, Kutoyis saw the people corralling bison, but went into a lodge occupied by old women. They suggested he go to a man's lodge, but he replied, 'No, I don't care for men's lodges, I am an old woman's child.' The women fed him, and he asked why they gave him only gut fat. 'Oh, my son', one explained, 'those bears might hear you. They take all the choicest parts. They take the wives of the people here away by force.'

Kutoyis went up to the pound, killed the fattest cow, and butchered it. The bears' cub came up to get the choicest parts for his family. Kutoyis struck the cub across the face, so that it ran home crying. The father and mother bears went up to the pound, stretching. Kutoyis jumped at the bears and stabbed each, killing it. He then went down to the bears' lodge and stabbed all the bears except one female, to whom he said, 'There can be more bears from you in the future.' Kutoyis then presented the bear-painted lodge [a medicine object] to the old women who had fed him, and invited the men to reunite with their wives.

Travelling higher up the river, Kutoyis encountered another camp, and again went into an old women's lodge. In this camp, a family of rattlesnakes exploited the people. Kutoyis drove bison into this camp's corral and distributed choice meat to the old women, then went into the snakes' tipi. Captive women in the lodge warned him to go out, but Kutoyis took the berry-flavoured drink prepared for the chief snake, who was sleeping, and drank it. Struck by Kutoyis's flint knife,[3] the chief snake awoke and reared up, whereupon Kutoyis cut off his head. Kutoyis then killed all the other snakes but one female, whom he allowed to crawl out to repopulate her tribe. He called the men of the camp to come to their wives, and gave the snake-painted lodge to the old women.

As he was leaving to visit farther, Kutoyis was warned by the women to avoid the Wind-sucker on the trail ahead. 'Don't go by him on the west side, he will suck you in. Go by him on the east side.' Kutoyis fearlessly jumped into the mouth of the grunting monster and in its belly saw many persons, some dead, others still alive. 'We shall dance', he told them. Putting his flint knife on top of his head like a topknot, Kutoyis danced jumping straight up so that his knife cut the monster's heart hanging above them. Kutoyis cut open the monster between the ribs and let out the surviving people.

Returning to the old women's lodge for a meal, Kutoyis was warned, 'Lower down the river there is a woman who will say to you, "Come here, let us slide for a while." That woman kills people.' As they predicted, Kutoyis was invited by the woman to slide with her. 'You must slide first', he said, and as she did, he threw out his flint knife, cut her rope, and she fell into a whirlpool below, dying.

Continuing on his travels, Kutoyis met a young man who invited him to wrestle. Kutoyis noticed a flint knife sticking a little out of the ground. Scuffling with the young man, Kutoyis threw him on his back against the knife and killed him.

Next, Kutoyis met a woman who said, 'Young man over there, come here for a while, let us play at "Sioux women." Stand over there. I shall throw my ball here, over.' 'No', said Kutoyis, 'I shall throw it first', and he threw it, smashing her head. Kutoyis again returned to the old women's lodge, and said to them, 'Old women warmed by fire, I killed all those, that treated you badly. Now I shall go south.' A young man wished to accompany him.

After travelling south[4] for a long time, the comrades came to a big lake. 'This lake is very dangerous. Here is another place where many of us have died', his companion told Kutoyis. 'Over there it looks black', replied Kutoyis, 'Let us go there.' 'Comrade, you must not stir this one lying there, the Blood-sucker. This is his, this lake.' Kutoyis declared, 'Comrade, I shall kill him, because he kills you all', and touched the Blood-sucker with his flint knife. The monster started to crawl. It was a long time before he got into the water. Kutoyis said to his comrade, 'I shall go into the water with him. Just stay now here. Look, when this water turns to blood, then I am killed, and then you must go to that hill. Stay there. You will see a cloud in the skies, the Thunder will help me. He will throw this one that killed me, out. And he will scatter this lake. It will never be water again. Then you must go from there.' So it happened.

Kutoyis is the same one who caused that there are no rattlesnakes and bears in this country [the Blackfeet Reservation]. They fled to the mountains, those that Kutoyis let live. He killed the Wind-sucker, the Wrestler, and the Slider.

Blood-Clot Boy is a motif found in Western North America, the Pacific, and Africa. The Blackfoot understand him to be a bison foetus aborted from one of the cows run into the pound by the old man. As a bison, Kutoyis would be more powerful than humans (the Horn Society's name refers to bison horns) and also favourably disposed towards them: bison are the staff of life, prosperity embodied, as symbolised in the little *iniskim* bison fetishes. A simple reading of the Blackfoot myth gives us a bison as saviour, transformed into human shape in response to the bison witnessing the unnatural horrifying behaviour of the inhumane son-in-law. The avenging bison then goes out to rectify similar abuses of the proper relationships between beings.

Taken in a broader perspective, the myth of Kutoyis invokes a series of cosmic manifestations. On the Northern Plains, bear is probably the equivalent of the Mesoamerican Jaguar Lord of the Earth (Hall 1989: 246), and thus equivalent also to Eurasian feline Lords of the Earth. Rattlesnake is certainly the Northern Plains version of the Mesoamerican Serpent, (Lord of) the Underworld, which in (Siouan) Hidatsa mythology is the husband of Grandmother Who Never Dies, the Earth who fosters the Corn Maidens and the sky-born hero Old Woman's Grandchild in her microcosmic earthlodge (Kehoe 1983). Wind-sucker seems to represent a tornado (cf. to the Oglala Iya, 'an enormous animal . . . [w]ith every breath he drew, trees and grass would all bend toward him' [Beckwith 1930: 434]).[5] The Slider is an otter, one of the anomalous water mammals believed to be more powerful than humans. The Wrestler may be a homosexual version of the vagina dentata (anus dentatus?); the more Americanised versions of Blood-Clot, those of Grinnell and Wissler and Duvall, give the Wrestler as a woman, but the direct Blackfoot versions of Black Boy and Uhlenbeck's narrator Joseph Tatsey make the figure a man. The 'Sioux women' game episode is blatantly sexual, the Blackfoot understanding a double entendre when the narrator speaks of a woman throwing balls: Sioux women are supposed, by the Blackfoot, to be promiscuous, and the game is also called 'Cree women', again in reference to supposed lust and promiscuity of women of alien groups. The final and truly fearsome monster is the Blackfoot version of the Siouan Underwater Panther that lurks in bodies of water, roiling them with its tail to make storms and seizing people who come down to the shore alone, especially at dusk. This serpent-tailed monster seems to be a manifestation of the Underworld Serpent, as Grandmother Who Never Dies's husband appears to be in Hidatsa mythology. The final episode of the Blackfoot myth climaxes in Thunder wreaking its power upon this, its counterforce; Bison, the bounty of the earth, allied with the People Above to overwhelm evil slithering within the chaotic waters.

The myth of Blood-Clot can be read as a cosmic contest between Good Fortune and Misfortune, a drama in two acts of four scenes each (four is the ritual number of the Blackfoot). In the first act, the Hero successfully battles the evil's oppressing communities; in the second act, the Hero pits his cunning against the terrors of the lonely places. At the end, formless Evil is destroyed, but the Hero, too, is consumed by the struggle. Such a conclusion reflects the Blackfoot view of the world as phenomena of power, not a battleground of titans but constantly shifting,[6] every manifestation of power relative in strength to every other. Seeking to 'be pitied', to be favoured with the granting of power by the Almightly through Its manifestations so as to live long in health and prosperity, Blackfoot beseech Manitou. They plead also with those of their fellows whose self-confidence and good fortune bespeak an uncommon power — contemporaries who in this way resemble the legendary Kutoyis — to exercise spiritual power in the supplicant's behalf. Everyday life becomes a series of contests of power, the presentation of self as more or less powerful than others, allying and realigning, challenging men and women to dare to recount their claims of deeds of valour, to dare to invoke power in healing or in sorcery.

The theme of contests of power has not been a focus of previous descriptions of the Blackfoot. It is implicit in the preoccupation with 'counting coup' (publicly reciting raiding and war exploits, and also women asserting their 'chastity' [Lewis 1941: 183]) that pervades traditional ethnographies of the Blackfoot. Perhaps the word 'contests' is not quite the best, for it is not always an agonistic confrontation as recounted in the myth of Kutoyis; rather, often it is *testing* power. In this context of cross-cultural comparisons, I have emphasised instances of *contesting*.

NOTES

1 I am deeply grateful to Ruth M. Lewis for her generosity in sending me a copy of the 1939 fieldnotes of herself and her husband.

2 I am indebted to Hugh A. Dempsey (personal communication, 22 October 1981) for the information that *kutoyis* means menstrual blood and the placenta, and that a different word is used for clots of circulating blood.

3 The repeated references to Kutoyis's flint knife, particularly the Wind-sucker episode in which it sticks up on top of his head, probably link the Blackfoot myth to Mesoamerican and Siouan mythology featuring the birth of Flint Knife

(Nahuatl *tecpatl*) which begat another generation of gods. The Iroquois Tawiskaron, Cherokee Tawiskala, Aztec Tlahuizcalpantecuhtli are represented as flint knives, although Tlahuizcalpan ('house of the rising sun') tecuhtli ('lord') is also Morning Star. Tawiskaron/Tawiskala is associated with ice, cold, and winter. Robert Hall (1983) discusses the many variations of Flint. I am indebted to David H. Kelley (personal communication, 16 February 1989) for additional commentary on the Mesoamerican and Siouan figures. An additional interesting occurrence is Dumézil's (1970: 161) note that 'the idol of the Lappish Þórr [Thor] had a piece of flint fixed in its head by a nail', a Sami borrowing of the Scandinavian myth of the duel between Hrungnir and Þórr.

4 South is the direction in which the spirits of the dead journey (Ewers 1958: 41).

5 I am indebted to Robert L. Hall for this comparison. Hall also notes (personal communication, 30 November 1989) the similarity between Kutoyis cutting his way out of Windsucker, with its belly full of people, and womb imagery.

6 Although their categories reflect an uncritical imposition of standard Western psychological labels, the Spindlers' (1965) interview data with Blackfoot reveal this constant weighing of potential and consciousness of the probability of its shifting. The Spindlers tag this 'literality' and say it limits 'conditional and conjectural thinking' but they sum up the characteristic thinking as 'what is, is; *what will be, no man knows*', which can express uncertainty on future relative power relations (Spindler and Spindler 1965: 19, my italics). A more sophisticated analysis of contemporary Blackfoot values and beliefs by the Spindlers's student McFee brings out somewhat more visibly the contingency factor that is high in Blackfoot consciousness (McFee 1972: 98–9, 101, 116–7).

REFERENCES

Beckwith, Martha Warren (1930). Mythology of the Oglala Dakota. *Journal of American Folklore* 43/170, 377–98.

Black Boy, Cecile (1973). Blackfeet Tipi Legends. In *Painted Tipis by Contemporary Plains Indians Artists*. Washington: U.S. Dept of the Interior Indian Arts and Crafts Board.

Dumézil, Georges (1970). *The Destiny of the Warrior*, trans. from the French by Alf Hiltebeitel. Chicago: University of Chicago Press.

Ewers, John C. (1958). *The Blackfeet: Raiders on the Northwestern Plains*. Norman: University of Oklahoma Press.

Grinnell, George Bird (1892). *Blackfoot Lodge Tales*. New York: Charles Scribner's Sons.

Hall, Robert L. (1983). A Pan-Continental Perspective on Red Ocher and Glacial Kame Ceremonialism. In *Lulu Linear Punctated: Essays in Honor of George Irving Quimby*, eds Robert C. Dunnell and Donald K. Grayson, pp. 74–107. Ann Arbor: University of Michigan, Museum of Anthropology Anthropological Papers No. 72.

—— (1989). The Cultural Background of Mississippian Symbolism. In *The Southeastern Ceremonial Complex: Artifacts and Analysis*, ed. Patricia Galloway, pp. 239–78. Lincoln: University of Nebraska Press.

Kehoe, Alice B. (1983). Ethnoastronomy on the American Plains. In *Ethnoastronomy in the Americas*, eds Von Del Chamberlain and Mary Jane Young. Proceedings of conference of the same name. Smithsonian Institution, Washington DC. 1983.

Lewis, Oscar (1941). Manly-Hearted Women among the North Piegan. *American Anthropologist* 43/2, 173–87.

Lewis, Oscar and Ruth M. Lewis (1939). Field Notes, Brocket Reserve. Manuscript in possession of Ruth M. Lewis; notes taken as members of Ruth Benedict's Columbia University field project.

McFee, Malcolm (1972). *Modern Blackfeet: Montanans on a Reservation*. New York: Holt, Rinehart and Winston.

Spindler, George and Louise Spindler (1965). The Instrumental Activities Inventory: A Technique for the Study of the Psychology of Acculturation. *Southwestern Journal of Anthropology* 21/1, 1–23.

Thompson, Stith (1957). *Motif-Index of Folk-Literature*. Bloomington: Indiana University Press.

Uhlenbeck, C. C. (1911). *Original Blackfoot Texts*. Verhandelingen der Koninklijke Akademie van Westenschappen te Amsterdam, Afdeeling Letterkunde n.r. deel XII, no. 1.

Wissler, Clark (1911). The Social Life of the Blackfoot Indians. *American Museum of Natural History Anthropological Papers* VII, Pt I, 1–64.

—— (1912). Ceremonial Bundles of the Blackfoot Indians. *American Museum of Natural History Anthropological Papers* VII, Pt 2, 65–289.

—— (1936). Star Legends Among the American Indians. *American Museum of Natural History Guide Leaflet* Series 91.

Wissler, Clark and D. C. Duvall (1908). Mythology of the Blackfoot Indians. *American Museum of Natural History Anthropological Papers* II, Pt 1, 1–163.

Contests in Etruscan and Roman Funerary Art

This chapter looks at the various images of contests that were used in the context of Classical funerary art: both the representation of actual funeral games, held as part of the funeral celebrations, and other contests which appear on funerary monuments with a more symbolic or allegorical meaning. Although the chapter deals mainly with Etruscan and Roman art, the most famous of funerals in the Classical world, and the most influential, was that of Patroklos as told by Homer in the *Iliad* (Book XXIII), and I shall start with a brief analysis of what happened there.

The ceremonies carried out by the Greeks in connection with Patroklos's death were prolonged and consisted of several parts: the lament; washing the body; the funeral feast with the slaughter of many animals, their blood being poured round the corpses; a procession with Patroklos being carried by his own men; the building of the pyre; the gifts of honey and oil. Then came the slaughter of four of Patroklos's horses and two of his nine dogs, presumably so they could accompany him to the Underworld (Vermeule 1979: 58–9; Griffin 1980: 3). Most spectacular and distasteful was the sacrifice of twelve noble Trojans, prisoners of war, by slitting their throats (more bloodshed): this was something Achilles had promised to do in revenge for the death of Patroklos, and it is made clear that it was an evil deed (*Iliad*, XXIII: 176: κακα δε φρεσι μηδετο εργα). The idea of human sacrifice at the tomb, though considered shocking, may not have been invented by Homer: Herodotus records that the Scythians of Thrace performed human sacrifices at funerals (IV, 71), though there it is not prisoners who are sacrificed but people whose function is similar to that of the dogs and horses, namely to serve the deceased. The killing of the Trojan prisoners was to be illustrated many centuries later by an Etruscan artist of the Hellenistic period, in the François tomb at Vulci, though it was there

Etruscanised, with Etruscan death demons present (Pallottino 1952: 115–24).

We are told, therefore, of three instances of bloodshed in the events commemorating Patroklos, with different reasons for them. In the first case, animals are killed to provide the meat for the funeral feast, though the blood itself is poured round the corpse; in the second, it is the particular animals loved by the deceased that are sacrificed, presumably so that they can accompany him; in the third case, human prisoners are immolated for revenge. In the *Odyssey* (XI: 23–50) it is blood that attracts and nourishes the hungry ghosts at the entrance to the Underworld. The association of the funeral with bloodshed was to continue, and we find instances of it even when society had become ostensibly more sophisticated, as late as the second century AD, in the letters of the most urbane Pliny the Younger: he retails with disgust how a father mourns to excess his dead son, slaughtering his pet Gallic ponies, dogs, nightingales, parrots, and blackbirds round the pyre (Book 4, Letter 2), but he elsewhere suggests, apparently approvingly, that gladiatorial games would be a suitable tribute for the funeral of his correspondent's wife (Book 6, Letter 34).

In later tradition, such as the gladiatorial shows held in association with funerals in the Roman world, bloodshed and contest were combined, but at the funeral of Patroklos they are separate elements, performing different functions. After the slaughter of the Trojan captives comes the cremation of the body, the collection of the ashes, and the erection of the tomb mound. It is only then that the funeral games are held, with valuable prizes to be competed for in a variety of events, none of which involved a battle to the death. By modern standards some of the umpire's decisions seem rather arbitrary; in fact, there are no true winners or losers, since nearly every competitor gets a prize, and indeed even non-competitors may be awarded one. Prizes are not allotted strictly on the basis of who wins in the contest but go to those who most deserve them. It is not made clear what purpose the games were designed to serve. Are they to please or benefit the deceased, or to commemorate him in a suitably grandiose way? Do they act as an activity to divert the living from their grief? Are they a means to distribute the deceased's belongings appropriately among the survivors? It is possible to think of several reasons why such a custom might have existed, and the significance that it may have had.[1] It is not certain to what extent the games were normal practice, but there are a few illustrations on both Mycenean and Geometric vases which may represent funeral games, and these may indicate that games could be held at funerals on both sides of the Dark Ages (Vermeule 1979: 62; Kurtz and Boardman 1971: 60, 187).

In the funeral games held for Patroklos the first event was a chariot-race, described in considerable detail. The second event was a boxing match: the challenge by Epeius shows that this could well end in death, though in the event his opponent was knocked out, but not killed. The next contest was wrestling, in which the contestants were evenly matched, so there was no outright winner. Then came the foot-race, an armed duel (in which the aim was to draw blood), discus-throwing, archery, and finally, throwing the javelin.

It is difficult to know whether such games continued to be a regular feature of funerary celebrations in Greece: they may have been reserved for those acknowledged as heroes (Kurtz and Boardman 1971: 202–3; Malten 1923–24: 316–7). Such a custom is, by its very nature, more likely to flourish in an aristocratic society, as a form of ostentatious commemoration of their dead by ruling families. Nevertheless, Aristotle refers to games held annually in democratic Athens in honour of her war-dead (*AP* 58: 1; see also Kurtz and Boardman 1971: 121). The practice certainly has not left much trace in Greek funerary art, though significantly it does appear later and on the periphery of the Greek world, as in the 'royal' tombs of Vergina in Macedonia, one of which (the 'prince's tomb') is decorated with a chariot-race frieze which possibly alludes to funeral games (Andronicos 1984: 204–6), and in the tomb of a local ruler at Kazanluk in Bulgaria (then Thrace) (Vassiliev 1960), which also has a chariot-race in the cupola. Funeral games were certainly held for Hephaisteion and Alexander himself (Arrian, *Anabasis*, VII: 14, 10), and for the father of King Nikokles of Cyprus (Isocrates IX: 1): a revival of funerary games is only to be expected with the resurgence of monarchy in the fourth century BC.

For a much richer artistic tradition we have to go to Italy, to the aristocratic societies of the South Italian Samnites and Lucanians, and the Etruscans in the North. An early tomb to be decorated with this theme is the Etruscan Tomb of the Augurs at Tarquinia, *c.*530 BC (Poulsen 1922: 10–4; Pallottino 1952: 37–42; Brendel 1978: 168–71). The tomb was named after a scene in the chamber originally interpreted as showing men about to take the auguries, but now recognised as umpires for the funeral games, and specifically for the wrestling match which takes up the central section of the wall. Two wrestlers face each other, just beginning to grapple: in between them, but intended to be understood as in the background, is a pile of metal bowls, the prizes for which they

Figure 1 Tomb of the Augurs, Tarquinia. Detail of the Phersu group. Photo: DAI neg. 81-4251.

are competing. On the opposite wall is a pair of boxers. If this were the only kind of contest illustrated in the tomb we might be tempted to think that this was an illustration of the funeral games of Patroklos rather than a real contemporary event, but along the wall from the wrestlers are figures involved in an incident that has no part in the Homeric funeral games (Figure 1). A large man in a distinctive costume and a mask with a long beard has another couple of contestants under his control: one is a man with his head enveloped in a bag so he is unable to see, and the other is an animal (probably a dog), biting the hooded man's leg. The hooded man has a club in his right hand, but obviously cannot hit the dog with it until he knows where it is, and in fact he is already bleeding copiously from several wounds. The situation is complicated by the fact that the man appears to be on a long lead which winds round his legs and therefore may trip him up — the other end of the lead is held by the masked figure. The same masked figure appears again on the opposite wall of the tomb, apparently running way, his pursuer possibly the man with the club, who has freed himself from the hood and the dog (Poulsen 1922: 12, Figure 5). Both the masked and the hooded figures appear on another tomb at Tarquinia, the Tomba Cardarelli (Moretti 1970: 93–102), and the masked figure is seen alone in the Tomba del Pulcinella (Poulsen 1922: 12–3, Figure 6): both tombs are of approximately the same date as the Tomb of the Augurs. The masked figure in the Tomb of the Augurs is labelled 'Phersu', a word which is said to be the origin of the Latin *persona* — he may be a human actor impersonating an Underworld character ('spooks on the loose' [Brendel 1978: 169]), though the contest itself looks very real, and not an act (despite Brendel's description [1978: 168–9] of the event as 'sinister pranks, much like clowns in a modern circus, as a side show between the athletic contests').[2] It is not certain exactly what is going on in these scenes, but they clearly reflect a specifically Etruscan contest, not part of the repertoire described by Homer. The event may also contain elements which were to be developed later in Italy. It is not just a trial of strength and skill but was designed to draw blood, possibly ending in death, and involved fighting with an animal. It is possible, too, that the contest used criminals or slaves (or other expendables) and/or professionals who put on a show, unlike the aristocratic amateur competitors at the funeral games of Patroklos.

Phersu, the hooded man, and the dog were not, however, standard characters in Etruscan tombs. Other tombs at Tarquinia and Chiusi show a variety of sports and their spectators, and suggest an atmosphere more akin to the Greek palaestra. The

Tomb of the Monkey at Chiusi (c.480–470 BC), for example, has
wrestlers with their adjudicator, a pair of boxers separated by a
stool on which a garment has been carelessly thrown, a javelin
thrower, chariots, and riders in a horse-race (Pallottino 1952: 65–
6). Boxers also appear in the Cardarelli Tomb at Tarquinia,
flanking the entrance (Moretti 1970: 98–9), and complex horse-
races also appear in several tombs. One tomb at Tarquinia is so
rich in its scenes of athletic events it is called the Tomb of the
Olympic Games (Moretti 1970: 103–20), though there is no reason
to suppose the games are other than the funeral contests. These
include a foot-race, a discus-thrower, jumping with weights,
boxing, and the chariot-race. In this scene one of the chariots has
come to grief, an ingredient which was to be standard in Roman
versions of the chariot-race, but which appears to have been
invented by the Etruscans, possibly even by the painter of this
tomb (Brendel 1978: 268). Another tomb at Tarquinia is known as
the Tomb of the Chariots, after its scenes showing preparations for
the chariot-race and the preliminary chariot procession, but there
are also preparations for other events involving athletes and
including a small armed figure who is thought to be an armed
dancer, a reminder that not all events in the games were
necessarily primarily athletic or competitive (Pallottino 1952: 61–
4). The dividing line is not always clear between what we might
recognise as 'contests' and 'displays'. The same is true of the
juggling scene from the Tomb of the Jugglers (Moretti 1970: 22–
3). A woman balances a tall object on her head, accompanied by a
flute-player; in front stands a man holding out a ring as if he is
trying to throw it on to the object. The scene is watched by a
seated man on the right. The Tomb of the Monkey has a very
similar scene with a single spectator, this time a woman dressed in
black, seated on a throne with a footstool and a sunshade (Brendel
1978: 275, Figure 191). It has been suggested that these are the
dead, watching and benefiting from their own funerary games
(Pallottino 1952: 65–6; Jannot 1987: 287), but equally the woman
has also been identified as the widow of the deceased, who would
have a prominent position at his funeral (Poulsen 1922: 26;
Brendel 1978: 275). In the Tomb of the Chariots, however, there is
less of a problem about the identity of the spectators. We are
shown a stand on which sit a crowd of aristocrats (men and
women), while below their servants or the lower classes recline on
the ground (Brendel 1978: 267, Figure 181). They all gesticulate,
giving a lively picture of the real and living crowd that might attend
and enjoy the funeral games.

There is nothing in the various contest scenes to confirm the exact circumstances in which the games were held, but it is generally accepted that they are pictorial records of actual funeral games held to celebrate the interment of members of the nobility, in conjunction with other events such as feasting and dancing (which are also represented in the tombs). The paintings then remain in the tomb as a permanent record for the deceased of the splendour of their funeral. The tomb chamber itself is perhaps seen as the threshold of the realm of the dead, since in some cases (such as the Tomb of the Augurs) we see a painted *trompe-l'oeil* doorway in the end wall: the games painted in the chamber therefore take place outside the door, that is, outside the tomb, and in the realm of the living.

Funeral games are popular on Etruscan tombs of the later Archaic and early Classical periods, c.530–450 BC and they represent the fullest expression of the theme. These tombs are aristocratic: it is easy to imagine the leading families of the Etruscan cities at their height vying with one another to provide ever more splendid and lavish games at their funerals. The idea of funeral games may have been exported by the Etruscans to Southern Italy, where it found fertile ground in another aristocratic society. It is now generally accepted that it was from this origin that the Roman gladiatorial games were to develop (Salmon 1967: 60; Frederiksen 1984: 339 n.33). Again we see some hint of this in the tomb paintings, particularly those of Paestum, dating to the later fourth and early third centuries BC (Salmon 1967: 60–1, 142; Sestieri 1956, 1958). A popular motif was chariot-racing, painted in a lively way with economical use of colour and shading: the chariots are clearly taking part in games, but not necessarily funeral games, though that is a strong possibility, given their location. Other scenes showing men fighting, armed with spears and protected by shields and helmets (Malten 1923–4: 325–8, Figures 15–18; Sestieri 1956: Figures 6, 13, 14, 21, 22, Plate 1; 1958: Figure 10) may or may not be part of funeral games: given the warlike nature of this society, with the aristocratic males generally appearing as mounted warriors in their tombs, such scenes *may* rather reflect their manly pursuits while alive, but we do also find boxing scenes, and these are more likely to be part of organised games (Frederiksen 1984: 145, 155 n.115). In the fighting scenes there is a certain emphasis on bloodshed: it was armed duels of this sort, originally organised as funeral games, which developed into gladiatorial contests organised for entertainment alone.

At Rome in the Republican period gladiatorial games were held as part of the funeral celebrations of the most important families

(Grant 1971: 18–9; Hopkins 1983: 3–7) — an early instance was at
the funeral of Brutus Pera in 264 BC. (Livy *Epit.* 16; see also
Salmon 1967: 60), but the practice continued well into the Imperial
period. The Romans used war captives, destined to die anyway,
criminals, and the least valuable slaves, trained to put on a good
show and fighting desperately for their lives. Even when the
gladiatorial games had become divorced from their funerary
origins, and were held primarily as a spectacle to amuse the
crowds, some indication of their origins was retained in the figure
of the slave dressed as the Etruscan death-demon Charun, who
removed the bodies from the arena (Grant 1971: 15). It is perhaps
surprising therefore that the Romans did not take up gladiators as
a major image in their funerary art. Gladiators do appear on a
small number of Central and Southern Italian tombs of the late
Republic/early Empire: a particularly elaborate version, with the
sponsor of the games seated in the pediment among the spectators
and the various combats on the frieze, was built in the vicinity of
modern Chieti and commemorated a certain Lusius Storax; while
another is the tomb of Umbricius Scaurus in Pompeii. Trimalchio,
too, in the fictional *Satyricon* of Petronius (*Cena Trimalchionis*,
Book 71) wants to have a representation of the fights of the famous
gladiator Petraites in his tomb. But gladiators do not commonly
appear on funerary monuments of Rome itself, despite their
traditional association with funerals, and despite their popularity
as decorative motifs for more mundane objects (such as lamps).
This is perhaps most surprising in the case of the long series of ash
chests and sarcophagi.

Gladiators there may not be, but other types of contest abound
in Roman funerary art. Perhaps the most obvious is the battle
scene, best known in the series of mid-Empire battle sarcophagi
modelled on official state monuments (Koch and Sichtermann
1982: 90–2). By showing the deceased in the thick of battle the
scene recalls his 'virtus', one of the Roman virtues. The same
principle appears to be behind the representation of hunting
scenes on Roman sarcophagi, popular at a slightly later date (Koch
and Sichtermann 1982: 92–7). The hunt or battle might also appear
as part of a mythological scene on a sarcophagus, such as the boar
hunt represented as part of the story of Hippolytus and Phaedra:
this, of course, was far from being a Roman invention. The
Etruscans of the Hellenistic period put a wide variety of myths on
their cinerary urns, and a large proportion of them involved
contests, often bloody ones. The duel between Eteocles and
Polyneices was a particular favourite, and another popular myth
was that of the chariot-race between Pelops and Oenomaus. It

may simply be that many Classical myths do have an element of violent contest in them, but certainly this aspect of the myths does seem to have been considered especially appropriate for funerary monuments by Etruscan and Roman artists and their patrons. One reason may have been a lingering belief that blood was an appropriate offering to the dead, a distant memory of more direct blood sacrifices at the tomb: another reason may have been an allegorical reading of myths to show that the principle of fate that governs the lives of the mythological characters also governs the rest of us, so that death, whenever it comes, is as violent, unexpected, and inevitable as in the myths.

Contests do not occur only in mythological contexts in Roman funerary art: there are also other settings, involving both ordinary humans and the semi-mythical world of little cupids or small boys. Chariot-races were a popular motif, and in some cases the reason for its choice may have been simply to do with the deceased's interests while alive: this would appear to be true of a funerary plaque in the Vatican (Helbig 1963: 724–5, no. 1010). This shows the deceased suitably enlarged and frontally placed standing alongside a race recognisably taking place in the Circus Maximus in Rome. He was presumably a sponsor of chariot-racing, who either paid for games to be held, or was a *dominus factionis*, patron of one of the teams. The same explanation may apply to a painting in Tomb C of the cemetery under St Peter's Basilica in the Vatican (Mielsch *et al.* 1986: Figure 42 and Plate 6). Here there are remains of a charioteer dressed in blue wearing a wreath and holding a palm branch, showing that he was a victor: there are further traces that suggest there was originally also a painting of his rival, a charioteer of the green team. If so, each charioteer was represented beside a turning post painted his own colour (blue or green), and in the field between them there was a large palm branch, decorated with blue flowers. The painting belongs to the Hadrianic phase of the tomb. In a recent publication of the tomb (Mielsch *et al.* 1986: 46), it is suggested that the decoration may indicate that the builder of the tomb, Tullius Zetus, was a *dominus factionis* for the blues.

This essentially pragmatic explanation was rejected by F. Cumont for another monument which is decorated with a very similar motif, the cippus of T. Flavius Abascantus, now in Urbino. This monument launches his discussion of the 'triumph over death' theme in the appendix to his massive work on Roman funerary symbolism (Cumont 1942: 457–84, Plate XLV). This altar is decorated with four separate elements: in the pediment an eagle; in the main field at the top a so-called 'funerary banquet' scene; in

the centre the inscription telling us that the monument was set up
by his wife and that he was an imperial freedman who had worked
in the civil service ('*a cognitionibus*'); and, below, a depiction of a
four-horse chariot at the gallop. The charioteer holds a wreath in
his right hand and has a large palm branch over his shoulder. In a
row of tiny letters above we are told exactly who he is: Scorpus,
one of the most famous charioteers of the Flavian period, who
appears in the poems of Martial (IV, 67; V, 25; X, 50, 53, 74). The
four horses are also labelled. Considering the great popularity of
chariot-racing at Rome at all levels of society, the presence of this
scene does not seem to me to be at all surprising. It is possible that
Abascantus was in some way involved in the patronage of racing
— but he might simply have been a great fan, and his wife
therefore considered the image of his favourite team to be an
appropriate way of decorating his tomb. Today such a possibility
does not, I think, seem preposterous, but to Cumont, writing in
1942, it seemed impossible that a high-ranking civil servant should
have on his tombstone the equivalent of a jockey winning the
Grand National (Cumont 1942: 459), evidence of changes in
modern values rather than a helpful assessment of Roman
attitudes.

   Whatever the truth of this particular case, Cumont goes on to
discuss the image of victory and its significance in Roman funerary
art, analysing other monuments where reference back to the
deceased's life and interests is less plausible. Quite a popular
decoration for sarcophagi, especially those used for children, was
the circus race performed by cupids (Figure 2) (Koch and
Sichterman 1983: 210–1, Figure 245). These often incorporate the
motif of the chariot that has come to grief, which also appeared on
the Etruscan Tomb of the Olympic Games. While it is always
possible that those buried in such sarcophagi were budding
charioteers, cut down before they could prove their skills, a more
general interpretation seems more likely. Cumont points to the
religious character of the circus, which, according to him, had
acquired mystic connotations under Eastern influence (Cumont
1942: 460). The circus itself, he suggests, represents the world, and
those who triumph there become a kind of *kosmokrator*, their
victory associated with that of the Emperor. They are made
perpetually invincible by a gift of heaven, and to win a race is
evidence of a superhuman soul. The deceased has performed
similar tasks and is therefore worthy of similar heroisation.
Cumont takes this line of argument further to suggest that the
cupids who take part in such scenes allude to a concept inspired by
Plato: they are to be seen as the winged souls who take part in the

Figure 2 Child's sarcophagus in the Naples Archaeological Museum (6712). Cupids racing in the circus. Photo: DAI neg. 62–845.

Figure 3 Ash chest of C. Minicius Gelasinus, Blundell Collection, Liverpool Museum: front. Cupids wrestling in the palaestra. Photo: author's own.

race of the cosmos, with violent competition between them (hence the fallen chariot) (Cumont 1942: 461, 349). So, in the case of his original example, the monument to Flavius Abascantus, Cumont believes we are meant to see the particular victory of Scorpus as a general image of the concept of victory, and as an indication that Abascantus too has won a victory, a victory over death, and that his terrestrial career was triumphal. Like much of Cumont's interpretation of Roman funerary symbolism, I think this goes too far and outstrips the evidence: Cumont's analytic method tends to involve the use of abstruse religious and philosophical concepts that are unlikely to have affected popular art except in the most bastardised and simplified form, and this makes it difficult to accept his conclusions in their entirety. Nevertheless, his ideas are useful up to a point.

The chariot-race is not the only theme in Roman funerary art to display the concepts of contest and victory. Cupids and small boys also compete in the traditional contests of the palaestra, especially wrestling (Koch and Sichtermann 1983: 212, Figure 246, 124, Figure 128; Aravantinou 1982). An example is the ash chest in Liverpool illustrated in Figure 3. Here the wrestling match has hardly begun: the competitors face each other warily before attempting to grapple. The whole panoply of the palaestra is represented here — the herms, the musicians, and above all the prize tables with their wreaths and palm branches to be awarded to the victor. Other representations of these themes on sarcophagi are sometimes more explicit in showing the victorious athlete/charioteer: he stands in a prominent position, facing front, often a little larger than the other figures, holding his palm branch and putting a wreath on his head (Cumont 1942: Figures 98, 100, and 101, Plate XLVI, 2 and 3). In such cases it is not just the competition that matters, but the victory, and it is this that has led Cumont to his conclusions about the significance of the contest scenes. However, it should be remembered that in many cases the contest is *not* yet decided: the wrestling bout is yet to take place; the race is not yet over, and it is this that seems to me to be a basic problem in accepting Cumont's thesis as an interpretation of the contest motif in all its manifestations.

There is one other form of contest which Cumont mentions only briefly in a footnote, though it has since been examined by Philippe Bruneau, who takes essentially the same line as Cumont (Bruneau 1965). This is the cock-fight, a motif which was quite popular on ash chests and grave altars in the first and early second centuries AD, but less so on later sarcophagi. In many cases the scene was reduced to a standardised motif of two cocks facing one another,

Figure 4 Sarcophagus of Malia Titia, Ostia. Detail of right lunette on front. Cock-fight. Photo: author's own.

either menacing each other or contesting one of the symbols of victory, the palm branch and wreath. Sometimes the conclusion of the fight is known, with one cock looking proudly victorious and the other dejectedly defeated, but this is not always so. On two monuments the motif is expanded into a small scene set in the palaestra. On the ash chest in the Vatican (Helbig 1963: 743–4, no.1033; Bruneau 1965: 116, no.94) the fight is over and the victor is known. The two cocks are in the charge of small boys: the one on the left leaves the scene, his bird under his arm with its head drooping — it may even be dead. The other boy restrains his exuberant bird which holds a wreath in its claw and moves towards the prize table. On a sarcophagus in Ostia there are two scenes that convey much the same idea: here the cock-fight appears in the two fields above the garlands (Calza 1954). The scene on the left shows the winning bird holding a wreath in its claw, while the loser crouches down before him: in the background are two boys, one of them blowing a trumpet (possibly to herald the winner). In the right hand scene (Figure 4) it is again clear that the left-hand cock has won: it stalks off to the left with its owner, giving a haughty look back over its shoulder to its defeated rival, which stands with head bowed, while its owner flees in tears with a friend trying to comfort him. Here we have winners, but also losers, and as much emphasis is given to the losers as to the victors.

Cumont's interpretation of contest scenes requires several assumptions, not all of which are valid in all cases. First, he assumes that contest scenes are allegorical and do not refer to the past life and circumstances of the deceased. Secondly, he thinks it is the victor and victory that is the important element in these scenes, rather than the competition. Thirdly, he identifies the victor with the deceased, and as an extension to this he identifies cupids as the souls of the dead. Finally, he assumes that the victory alluded to must, because these are funerary monuments, be a victory over death, a victory which results in the heroisation of the deceased. Bruneau (1965: 116) and Aravantinou (1982: 82) agree with him, in outline at least.

Let us look at each of these assumptions in turn. First of all, there is no need to reject out of hand the idea that sometimes atheletes, charioteers, and their attributes were chosen for funerary monuments because the deceased had been in some way involved in these activities while alive: this may not be so in all cases, but it is a possibility in some. In the second case, we have seen that not all scenes put emphasis on the victor, and the element of competition must be seen as just as important in some cases as the ultimate victory appears to be in others. And in some scenes, such as the cock-fights, the loser is given as much emphasis as the winner, and who is to say which (if either) is to be identified with the deceased? This takes us on to the third assumption: I would agree that in some of the instances cited by Cumont the victor is given such prominence that his identification with the deceased is likely,[3] but this is by no means true of all contest scenes. As for the suggestion that the cupids represent souls of the dead, this is the part of Cumont's discussion which seems to be least thoroughly thought through, and which does not seem to accord very well with his other interpretations. The facile interpretation of cupids as representing souls is often made, but I find this a difficult interpretation to accept. Such figures — little boys with or without wings — are ubiquitous in Roman art, and are every bit as popular in non-funerary secular art as on sarcophagi. They live in a world of their own, aping the deeds of adults in coyly cute scenes such as the famous paintings from the House of the Vettii at Pompeii, among them a scene showing cupids racing very like those in funerary art. A scene which uses cupids instead of human beings is clearly non-realistic and belongs in a fantasy world: an allegorical meaning may be intended, but the cupids are not necessarily souls, nor do we have to assume that because they appear on a funerary monument they must have a deeper and more complex meaning than they would have in another context.

Finally, there is Cumont's assumption that the victory (where there is a victory) is a triumph *over* death. For this he gives no specific evidence: for him, apparently, the conclusion is self-evident; for me, it is not. Indeed, in another motif of struggle and contest it is possible that what we see is an illustration not of the victory over death, but of the victory *of* death. These are the scenes, popular in various periods, of animals, real or mythical, fighting each other, usually with an indication that one is stronger and destined to be the winner. If these have a symbolic meaning at all it is surely to recall the ravening power of death, rather than man's victory over death.

Clearly the image of the contest was a powerful and popular one in Roman funerary art, and could be manifested in many different forms. Winning and the victor may be important, but they are not necessarily represented, and some scenes put less emphasis on this aspect. As Cumont says, various types of contest played a large part in Roman society, and the competitors were often treated as if they were superhuman. The image of the world as the circus track, analogies between sport and everyday life, were common: but this is on the level of metaphor, and we do not have to look to mystic ideas from the East to understand it. In real-life contests some win and some lose: the individual is sometimes successful and sometimes not, but he is a hero as long as he keeps winning. The race, the wrestling match, and the cock-fight are ruled by chance or fate as much as life itself. It is this analogy, I believe, which explains the majority of the contest scenes rather than Cumont's emphasis on the heroised victor over death. Cumont's view may apply to those scenes produced in the later Empire which put emphasis on the victor, but it is significant that they belong to a period when salvation, either through the direct intervention of a deity or as a result of one's own worth, was hoped for or believed in by many. In earlier times there was less reliance on salvation, indeed less evidence that people were even very concerned about their fate after death. They lived in a world ruled by Tyche, Fate — a world in which you could only hope to win some and to lose some, and could not expect to control what happened to you (though you might try (Martin 1987). This is very much the Hellenistic world view, coming between an earlier belief in the efficacy of correct actions in relation to the gods and the later personal contract made with saviour deities. Life is like one of the races we saw on the sarcophagus in which even the best charioteer may suffer a fall; or like the wrestling bout where the competitors are evenly matched; or the cock-fight where if one wins, the other loses. Some of the Roman patrons who commissioned funerary monuments may have intended the scenes to mean more than this, but such an interpretation, it seems to me, better explains the general popularity of the contest scenes than does Cumont's 'victory over death'.

Contest scenes of various sorts were popular in both Etruscan and Roman art over a long period but their significance did not remain static and there can be no single explanation for them. The realistic representations of funeral games in Etruscan tombs are on one level simply a record of what went on at an aristocratic funeral: the fact that funeral games were held at a person's death was a sign of nobility and possibly also of heroic status, in either case a status symbol. A preference for the bloodier sports may also reflect the belief that the dead appreciated blood, and this idea may have continued to affect people subconsciously even in later, supposedly more sophisticated, eras. The realistic depiction of games continued into the early Empire, though now the deceased may have had a different relationship with the games concerned, as a sponsor, organiser, or fan. But at the same time realistic representations also began to give way in the Hellenistic period to scenes that appear to use the contest as a metaphor for the trials and tribulations of life itself, a life governed by chance, fate, and destiny. At first the metaphor did not place emphasis on the victory, or equate the deceased with the victor, but as time went on there seems to have been greater emphasis placed on the victor in the contest, so that in the later Roman Empire the theme could be used to suggest that the victor is someone who transcends the destiny of humankind and is heroised after death. Thus, throughout its Classical history, the contest theme appears to have been related to the concept of the hero, but there was a series of shifts in the relationship between them: at first, the funeral game is an appropriate form of celebration for a hero and denotes heroic status, but by the end it is the dead, metaphorically, who takes part in the games, and by his victory proves his heroism. This last encoding, however, was not as widespread as Cumont suggests, and we should be careful not to ascribe more complex meanings to the contest scenes than the evidence will support.

NOTES

1 Sansone's theory, that sport is a ritual sacrifice of physical energy, offers attractive possibilities (Sansone 1988); his parallel thesis relating this to hunting ritual (which in any case is less convincing) does not seem to be so relevant here.

2 Earlier writers tended to accept without question the connection between such bloody contests and human sacrifice: thus Poulsen (1922: 14) wrote, 'These combats were a piquant and exciting substitute for actual human sacrifices in honour of the deceased noble or the gods, and as one of the parties was given a chance to save his life the practice may even be considered an advance in humanity.' Sestieri (1958: 50) gives the same explanation for a gladiatorial scene on a Paestan tomb, It is tempting to see such a connection, but there is little positive evidence to support it.

3 One piece in particular suggests that such an identification was intended: the sarcophagus in Milan dedicated to a six-year-old girl, Octavia Paulina (Cumont 1942: Plate XLVI, 3; Aravantinou 1982, 82, Figure 9). Here the palaestra scene does not involve the usual boys only but instead the children are of both sexes and the centrally placed victor is female. Such a deliberate change in the sex of the figures must surely indicate the intention that the deceased be identified with the victorious athlete in this particular case. It need not, however, indicate that such an identification was commonplace.

REFERENCES

Andronicos, Manilos (1984). *Vergina. The Royal Tombs and the Ancient City.* Athens: Ekdotike Athenion S.A.

Aravantinou, Margherita Bonanno (1982). Un framento di sarcofago Romano con fanciulli atleti nei Musei Capitolini (contributo allo studio dei sarcofagi con scene di palaestra). *Bollettino d'Arte* 67/15, 67–84

Brendel, Otto (1978). *Etruscan Art.* Harmondsworth: Penguin.

Bruneau, Philippe (1965). Le motif des coqs affrontés dans l'imagerie antique. *Bulletin de Correspondance Hellénique* LXXXIX, 90–121.

Calza, Raissa (1954). Un nuovo sarcofago Ostiense. *Bollettino d'Arte* 39, 107–15.

Cumont, Franz (1942). *Recherches sur le symbolisme funéraire des Romains.* Paris: Librairie Orientaliste Paul Geuthner.

Frederiksen, Martin (1984). *Campania.* Rome: British School at Rome.

Grant, Michael (1971). *Gladiators.* Harmondsworth: Penguin.

Griffin, Jasper (1980). *Homer on Life and Death.* Oxford: Clarendon Press.

Helbig, Wolfgang (ed.) (1963). *Führer durch die öffentlichen Sammlungen klassischer Altertümer in Rom* I: *Die Päpstlichen Sammlungen im Vatikan und Lateran.* Tübingen: Ernst Wasmuth.

Hopkins, Keith (1983). *Death and Renewal.* Ch. 1: 'Murderous Games'. Cambridge: Cambridge University Press.

Jannot, Jean-René (1987). Sur la représentation étrusque des morts. In *La mort les morts et l'au-delà dans le monde romain* (Actes du Colloque de Caen 20–22 Novembre 1985), ed. F. Hinard, pp. 279–91. Caen: Université de Caen.

Koch, Guntram and Hellmut Sichtermann (1982). *Römische Sarkophage.* Munich: C. H. Beck'sche Verlagsbuchhandlung.

Kurtz, Donna and John Boardman (1971). *Greek Burial Customs.* London and Southampton: Thames and Hudson.

Malten, L. (1923–24). Leichenspiel und Totenkult. *Mitteilungen des Deutschen Archaeologischen Instituts, Römische Abteilung* XXXVIII–XL, 300–40.

Martin, Luther (1987). *Hellenistic Religions. An Introduction.* New York and Oxford: Oxford University Press.

Mielsch, H., H. von Hesberg, and K. Gaertner (1986). *Die Heidnische Nekropole unter St. Peter in Rom. Die Mausoleen A–D.* (Memorie of Atti della Pontificia Accademia Romana di Archaeologia serie III, vol. XVI, i). Rome: 'L'Erma' di Bretschneider.

Moretti, Mario (1970). *New Monuments of Etruscan Painting,* trans. D. King. London: Pennsylvania State University Press.

Pallottino, Massimo (1952). *Etruscan Painting.* Geneva: Albert Skira.

Poulsen, Frederik (1922). *Etruscan Tomb Paintings.* Oxford: Clarendon Press.

Salmon, E. T. (1967). *Samnium and the Samnites.* Cambridge: Cambridge University Press.

Sansone, David (1988). *Greek Athletics and the Genesis of Sport.* Berkeley, Los Angeles, and London: University of California Press.

Sestieri, P. C. (1956). Tombe dipinti di Paestum. *Rivista dell'Istituto Nazionale d'Archeologia e Storia dell'Arte* V, 65–110.

—— (1958). Tomba a camera d'età Lucana. *Bollettino d'Arte* 43, 46–63.

Vassiliev, Assen (1960). *The Ancient Tomb at Kazanluk.* Sofia: Bulgarski Houdozhnik Publishing House.

Vermeule, Emily (1979). *Aspects of Death in Early Greek Art and Poetry.* Berkeley, Los Angeles, and London: University of California Press.

GEARÓID Ó CRUALAOICH

Contest in the Cosmology and the Ritual of the Irish 'Merry Wake'

The merry wake is a phenomenon of Irish popular culture that flourished in the period from early modern times to the end of the last century. In English usage, in the context of popular culture, the term 'wake' usually refers to a-vigil-cum-festival marking the anniversary of the dedication of a church. In Irish usage, however, the term 'wake' has the primary meaning of a vigil in a corpse-house, i.e. a house where a recently deceased individual is 'laid out' for a period of two days and two nights during which time family, kin, and neighbours engage in customary mortuary rites that culminate in burial of the corpse. Such a 'wake' is described as 'merry' because of the elements of revelry and carnival that it might include, though not every wake was equally merry, as will become clear.

As a phenomenon of the popular or peasant culture of early modern Ireland, the merry wake was exotic in the eyes of travellers from other lands and scandalous in the opinions of both the landed gentry and the urban and urbanising middle class within Ireland. Sustained efforts were made by ecclesiastical authority throughout the early modern period to suppress the merry wake or to rid it of its chief components of ritual mourning and revelry.[1] These efforts met with little success until post-Famine times, when major social and socio-economic changes in Irish society had transformed Irish popular culture in profound ways that included the conceptual and devotional reinforcement of Roman Catholicism.[2] Until the middle of the nineteenth century, however, Irish rural communities — especially ones with substantial numbers of Irish speakers — had comprised a majority who combined the practice of Christianity with a traditional allegiance to ancestral cults derived from a cosmology and a metaphysics whose origins lay in Celtic religion.[3]

A structural principle of that traditional cosmology was the centring of community life into occasional sacred assemblies that functioned to renew social order as well as to celebrate and propitiate supernatural personages and powers. The ancient Irish royal assemblies at burial sites — *Tailtiú, Carmun*, etc. — with their athletic contests, their board-games, and their horse-racing are early examples. The merry wakes of more modern times combined a Christian perception of the passage of the soul of the deceased into a Christian afterlife with an equally viable perception of the transition of the deceased into the far more contingent otherworld dimension of the native otherworld realm — a realm where Celtic gods and goddesses lived on as the *slua si* ('fairy host') and were joined at death by human ancestors whose passage into this otherworld realm had to be carefully facilitated by the bereaved family and local community. Thus, the merry wake is a traditional assembly carrying sacred significance in both Christian and ancestral religious terms. It functions at the liminal interface of life and death in the community and is a prominent part of the means by which the community effects the re-establishment of social order in the face of the disruptive power of death.

As mortuary ritual with sacred significance, the merry wake is necessarily also a social institution that expresses the bereaved community's awareness of its social loss and the necessity of social adjustment so as to re-establish normal relations. As both sacred and social ritual, then, the merry wake lends itself to interpretation of the kind that both Hertz (1960) and van Gennep (1960) brought to the fore in the anthropological study of death. Pointing out what they call Hertz's 'crucial insight', Huntington and Metcalf (1979: 17) state:

> Close attention to the combined symbolic and sociological contexts of the corpse yields the most profound explanations regarding the meaning of death and life in almost any society.

Van Gennep's analysis of passage ritual (e.g. 1960) identified two universal features in the mortuary rites of traditional cultures: namely, public mourning, and a period of licence. In the mortuary ritual of pre-modern Irish popular culture, these two features are manifested as (*a*) the 'keening' tradition (Ir. *caoineadh*: 'crying', 'lamentation'), involving the services of specialist semi-professional, female mourners; and (*b*) the 'play' tradition, whereby performance of story-telling and of elaborate organised games and amusements together with disorderly and abusive behaviours are sanctioned in the corpse-house for the duration of the wake.

The *bean chaointe*, the keening woman, retained by the family of the deceased to cry over the corpse and to compose extempore verse that both praises the deceased and expresses grief at his or her demise, is both a symbol and an agent of the transition of the individual deceased to an otherworld that is ambiguously Christian and ancestral/Celtic/fairy. A communal response to the dissolution and renewal of social order in this life in the face of the intrusive contact from the otherworld which the death of the individual constitutes is symbolised in the person of the *cleasaí*, or 'borekeen', the male, jester-like figure who acts as a kind of master-of-ceremonies for the disruptive rowdiness of the wake-games. This pair, the *bean chaointe* and the 'borekeen', can be said to represent the widest sense in which the merry wake itself constitutes a contest. The force of the *bean chaointe*, seen against the traditional cosmology of Irish popular culture, is to assert the hegemony of the supernatural in human affairs and to give powerful expression to the sovereignty in supernatural and social realms alike of the goddess figures who are pre-eminent in Celtic and Irish tradition.[4] The force of the 'borekeen', on the other hand, is to assert the continuing vitality and the potential for creative renewal of the human order and the individual men and women who live by it and who will survive this mortal contact between the supernatural powers of the otherworld and the field of human affairs. While not pitted against each other in direct competition as a part of the wake ritual, nevertheless the opposition of *bean chaointe* and 'borekeen' in cosmological terms is clear, and enables us to see the merry wake itself as a kind of contest in liminal time and space for control or dominance of life. Figure 1 attempts to show this in schematic form.

OTHERWORLD

Bean chaointe

Individual Transition/Incorporation into Otherworld

	SACRED	
WAKE	———	ASSEMBLY
	SOCIAL	

Communal Reversal/Regeneration within Social Order

'Borekeen'

HUMAN SOCIAL REALM

Figure 1

Within the behaviours of the merry wake there are other lesser contests that are acted out in various ways during the course of the combined mourning and merriment. Before going on to consider these and their cosmological significance, it is necessary to discuss the notion of 'timely' and 'untimely' death in Irish popular cultural tradition, since this distinction determines the extent to which merriment co-occurs with mourning in each specific instance of individual demise, and thus determines the character of the ritual occasion within which the various lesser contests can occur.

Basically what is at issue in the distinction between timely and untimely death in Irish tradition is the recognition of two separate cosmological mechanisms or agencies of death: the one ancestral or Celtic; the other Christian. A distinction made between the 'natural' and the 'magical' is also involved here, with untimely death, attributed to the agency of the powers and personages of the contingent ancestral/Celtic otherworld, being regarded as 'magical'; while timely death, attributed to the agency of the Christian divinity and his ordinance, is held to be 'natural'. Thus, an elderly person, deemed to have lived out an entire life-span and to have enjoyed the full range of human fulfilment in the course of that life, was thought to have undergone a natural death that was timely and that offered no very sudden or serious challenge to the social order. On the other hand, unforeseen or accidental sudden death, especially in those who were young or in the fullness of life, was both judged untimely and attributed to the magical agency of the fairies. Fairy abduction was said to be the cause of such sudden or accidental deaths as those of the young mother who dies in childbirth, the young man who, collecting firing, falls over a cliff-edge, the cowherd who perishes on the home mountain, the fisherman drowned in calm weather, the child who grows sickly and wastes away. Many legends existed that told of encounters by members of the community with people who had been abducted in this way, and such encounters were held to confirm for the community the continued existence of the suddenly or untimely deceased in that contingent fairy realm that was under the same sky as human society and that was the alternative otherworld in Irish tradition to the Christian concept of a timeless and invisible heaven beyond all human perception. By contrast with the unfathomable or transcendent Christian afterlife, the ancestral/Celtic fairy otherworld was envisaged as being located underground in the locality of the community — to be entered at the *lios* (ringfort), the hillside cavern, the lake shore, and in other specific local sites.[5]

The fairy abduction involved in untimely deaths was held to be not only a grievous rupture of social order but a serious challenge to the whole social realm and its continuing vitality. Revelry and merriment were subdued in the waking of the untimely dead by comparison with the behaviours indulged in during the wakes of the elderly and timely deceased (see Figure 2). This reduction of emphasis on the function of the 'borekeen' — mirroring the way untimely death is perceived as seriously impairing society's 'natural' ability to restore social order in the face of 'natural' death — placed increased emphasis in such cases on the function of the *bean chaointe*, whose role is chiefly concerned with effecting the transition of the deceased to the afterlife and with ensuring his or her incorporation with the family ancestors as these are perceived to co-exist dualistically both in a Christian heaven and in the fairy realm.

		UNTIMELY DEATH		
Explanation	*Agency*	*Response*	*Emphasis*	*Performer*
Magical	Fairy Interference	Restrained	Incorporation	*Bean chaointe*
Natural (Religious)	God's will	Unrestrained	Dissolution	'Borekeen'
		TIMELY DEATH		

Figure 2

Such extra concern with the proper incorporation of the deceased is justified in the sense that fairy abduction, the attributed cause of the deaths in question, has disrupted and impaired the process of incorporation both in the social realm of this life and in the afterlife otherworld realms. Richard Jenkins has already suggested that explanations of death as fairy abduction are tied to the process of kinship group incorporation in Irish tradition (Jenkins 1977). Faulty incorporation denies the individual access to the benefits of membership of his/her kin-group both in this life and in the next. This is particularly in question in the case of the death of young married women, in childbirth death, for instance. Such a victim of fairy interference in human affairs is singularly clearly not fully or finally incorporated into her husband's family, and her incorporation into an appropriate ancestor grouping in the otherworld is now problematic following on her abduction. In such cases, the ethnographic evidence we can come by in the extensive archive of oral tradition collected by the fieldworkers of the Irish Folklore Commission from 1935 suggests that a struggle or contest frequently takes place between the family of origin of the deceased and the family of her husband, each side seeking to bury the body in their ancestral graveyard with their own ancestors.

Irish tradition held that fairy abduction resulted in the substitution of a fairy 'changeling' for the human victim so that, in a sense, the corpse to be waked and buried following untimely death was not that of the human deceased at all but rather a fairy corpse. We may note that a known way is reported in the oral narrative tradition of West Muskerry whereby the changeling corpse can be disposed of so as to restore the abducted deceased to her human family.[6] This way of disposing of the changeling corpse involves opening the coffin three times, once on each of three bridges over which the funeral procession passes on its way to the woman's ancestral graveyard. On the third bridge, in the case where you are certain that it is not really your own kinswoman but a changeling who lies in the coffin, fling the corpse out of the coffin down into the running water and your own kinswoman will be at home before you when you come there.

In reality, it was very common for the two sides of the deceased's family to engage in ritualised battle to repossess the body of the deceased and bear it off for burial in their own side's burial ground, and I think we are justified in assuming that the majority of such ritual battles took place as part of the obsequies of the untimely dead — especially married females dying 'before their time', as it were. In such an instance, both sides of the dead woman's family will be especially concerned to bury the body in their ancestral graveyard as being the nearest they can get to the recovery of their dead kinswoman from the anonymous fairy host and her reincorporation with her kin group both in this world and in the next.

One such ritual battle is reported in exemplary fashion from the year 1797 by a Chevalier de Latocnaye, a native of Brittany, who was travelling then in the Killarney region of County Kerry in the Irish southwest. I quote John Stephenson's 1917 Belfast translation of de Latocnaye:

> I was witness here, a few days after, of a somewhat strange scene. Hearing the funeral bell, I went out to observe the procedure. It was the funeral of a poor woman who was being carried to her last resting-place, the coffin surrounded by a prodigious number of females who wept and chanted their 'hu

lu lu' in chorus, the men looking on rather indifferently. When the funeral arrived at the head of the 'T', that is at the end of the principal street of the town, a singular dispute occurred between the husband and the brother of the deceased. One of the parting ways led to the Abbey of Muckross, where it was the custom for the family of the husband to bury their dead; the other led to Aghadoe, where were buried the family of the brother. The latter assumed the right to direct the funeral towards Aghadoe, while the husband wished to go in the other direction to Muckross. The friends of the two parties took hold in turn of the remains of the poor woman, each wishing them to be carried to the side they favoured; but each finding themselves unable to succeed, by common accord they deposited the bier on the street and commenced a vigorous fight to determine by blows of sticks to which side the remains should be carried. I was at the time with the minister of the parish, Mr Herbert, who was also a Justice of the Peace. With great courage he threw himself into the middle of the fight, seized the collars of the two principle combatants, and, after some explanation, he decided that the husband had the right to decide where his wife should be buried. He allowed the husband then to go without letting go hold of the brother-in-law, and the funeral moved in the direction of Muckross. I remarked that neither fight nor controversy which followed arrested the cries of the wailing women, who continued to beat their breasts, tear their hair, and cry 'hu lu lu' as if neither fight nor controversy proceeded.

The Chevalier and the clergyman cum legal officer think they have witnessed a senseless outrage, a breakdown of law and order bordering on the obscene. To the anonymous actors in the event, however, its meaning is very different. In terms both of robust assembly and of ritual mourning, honour and respect have been shown publicly to the individual deceased by both sides of her family. Her radical transition to a new order is mirrored and marked in the public dramatic of the transfer of her mortal remains, a dramatic that also represents the re-establishment of a renewed social order in the aftermath of her demise. De Latocnaye finds it especially remarkable that the keening should continue throughout the fracas. But this is of the very essence of the merry wake/funeral, that it should simultaneously serve a dual function; mourning a transition and also resolving and removing social tension. He is also surprised when, as he says, 'the peasants showed the greatest respect to the magistrate, and submitted promptly to his decision'. They do so presumably because this is not a real battle and because they have already shown in their mounting of such a public ritual — in this case in the actual physical presence of their legal masters and social betters — their claim to independence of an imposed civil authority and their allegiance to ancestral life-ways.

I turn now to some of the other elements of contest that are incorporated into the ritual behaviours of the early modern Irish wake and funeral. There is in each of these a sense in which the struggle of the contest is as much a struggle to protect as it is a struggle to dominate and an underlying orientation of the whole ritual is that the social order needs protection against the dangers of contact with supernatural power which death inevitably brings with it — for the immediate family and kin of the deceased and also for the neighbours and the community in general. For instance, when a head of household died it was necessary as part of the process of 'laying out' the corpse to turn the body head to toe from the position in which it had expired. This was done to protect the deceased's family and all those about to assemble at the wake-house from the malevolent power of death which had just visited itself on the community — whether from fairy or Christian/'natural' cause.

We can note also in this respect that once death had occurred there was a strict taboo on any family member handling the corpse for the purpose of 'laying it out'. This task was performed by a female neighbour who was the local specialist in such work, washing the corpse and dressing it in its grave clothes, arranging its limbs, etc. The Gaelic name — very widespread — for this person was *an bean bhán*, 'the white woman', and she is, clearly, one of that band of female practitioners of various skills who are, as it were, licensed and symbolically insulated for contact with supernatural forces. The wise woman (*bean feasa*), the country midwife (*bean ghlúine*), the keener (*bean chaointe*) are others with a similar role, and they are all ultimately relatable in cultural logic to the goddess/hag figure that plays such an important role in medieval Gaelic mythology and its folklore reflex in both Ireland and Scotland.

While the *bean bhán* performed her duties in laying out the corpse, a contest took place between the deceased's agnatic and affinal relations which could last right through the wake until the corpse was coffined for burial. The point of the contest here was to 'turn the next death' onto the 'other' side of the family and it involved keen rivalry and even physical struggle to turn or reverse a variety of objects with which the deceased had contact in death. The pillows, sheets, and other bedclothing on which the deceased last lay were to be 'turned' in this fashion, as was the mattress or pallet, and even the bed itself, once the corpse was moved to the kitchen table which was the standard location of the corpse for the 'lying out' of the wake period. At the end of the wake a coffin was brought into the house and the corpse lifted into it from the table. The table itself then became a further object of rivalry and struggle between the two sides of the deceased's family — each party wishing to topple it so as to ward off from themselves the next subsequent mortal visitation of the community by supernatural forces. Incidentally, the time to elapse before such visitation could be roughly gauged by the condition of the corpse — in respect of how well or not it stiffened in death. A limp corpse was regarded as a certain sign of early impending death in the same household.

The funeral procession to the graveyard could itself be the occasion for further manifestations of rivalry and contest deriving from the belief that, at any given time, the last deceased individual buried in a graveyard was obliged to act as servant to all the others buried there until such time as another, newcomer deceased arrived as a replacement. Should two funeral parties meet on the way to or at the same graveyard, they would each strive to be the first to complete the burial — even struggling to impede each other, this despite a well-known saying that appeared to regulate such an eventuality: *tosach na reilige don óige* (youth has precedence into a graveyard!).

On the Cork–Kerry border north of Adrigole on the road to Tuosist, a rock called *Leac na gCorp*, the Corpse flagstone, marked the county boundary and funeral parties from either county bringing a body across for burial in the deceased's home ground frequently clashed here with mourners from the recipient county. The people of the recipient county required the coffin to be deposited on *Leac na gCorp*, from where they would then themselves take it onwards. The donating county party, however, inevitably attempted to press on into their neighbour's territory in the belief that to do so would carry bad luck and other misfortune out of their own country for that year. The fortunes of the succeeding harvest were understood to be at stake in this and frequent battle is reported from the area.

A different kind of struggle is reported as having frequently taken place in the course of the funeral procession and at the graveyard. This is the contest resulting from the active enmity towards the *bean chaointe* of the majority of parish clergymen. It appears that a priest of the parish in which the burial was to take place would often arrange to meet the funeral procession *en route* to the graveyard, or in some cases, would join in the procession in its later stages. The clergyman would take up the leading position in the procession, riding or walking in front of the bier, which would be borne on the shoulders of four men (in relays), or else — for longer funeral journeys — would be in a horse-cart. Whether shouldered or drawn, the coffin would be attended throughout by one or more 'keening women' [*mná* (pl.) *chaointe*] and, in the case of a coffin borne in a horse-drawn cart, the chief *bean chaointe* would ride in the cart, sitting on the coffin and leading the crying party of female mourners in the *olagón* as described above by de Latocnaye.

The priest meeting such a funeral was likely on occasion to confront the 'keening women' and to attempt to disperse them and drive them out of the funeral. A verbal duel frequently took place on such occasions between the clergyman and the *bean chaointe*, with insults and mockery and curses being exchanged. During the nineteenth century there were instances when, in such situations, physical violence was resorted to by the priest. Ó Súilleabháin (1967: 143) reports that his own father witnessed this happening at a funeral that took place about 1880, the priest using a horsewhip to repel his female opponents. At first glance this contest is about the imposition of male, clerical authority on areas of the social life of that sector of the Irish rural community hitherto beyond the domain of church control. Such extension of clerical social control is certainly at issue here, but we can see a deeper significance also in the confrontation if we remember that the *bean chaointe* is, in terms of the ancestral cosmology of the Gaelic world, a repre-sentative of the Mother Goddess and her cultural transforms — the Goddess of Sovereignty, the Goddess of War and Death and, in later folk tradition, the Supernatural Death Messenger.[7] This latter figure of the 'Banshee' (from *bean sí*, 'otherworld female'), whose crying is heard in the vicinity of the household about to experience the death of one of its members, is the nearest link the *bean chaointe* and the other flesh and blood females who are linked with her (all having a specialist role in relation to otherworld contact) have with the range of mythological females who figure in the ancestral/Celtic cosmology that continues to inform the ideological allegiance alternative to Christianity in Gaelic tradition. In the confrontation, then, of priest and *bean chaointe*, we can see opposed to each other office holders and, as it were, official representatives of the two alternative cosmologies that underlie the value and belief systems of Irish popular tradition down to modern times. Figure 3 attempts to set out this opposition in terms of the major personages and roles discussed above.

OTHERWORLD	HUMAN SOCIETY		OTHERWORLD	
War/Death Hag	*Bean bhán*	Hierarchy		
{Mother Banshee {Goddess	*Bean chaointe*		Saints	Male Divinity
Spouse Sovereignty/Fertility	*Bean ghlúine*	Clergy		
CELTIC/GAELIC			ROMAN/CHRISTIAN	

Figure 3

The organised games played at wakes included further manifestation of the tension or struggle between clerical authority and
an ancestral value system in Irish traditional culture. The clergy
themselves and their sacramental ministrations were travestied
and mocked in some of the imitative and miming wake-plays.
Mock-marriage and mock-confession organised as part of the
amusement under the direction of the 'borekeen' gave scope in
terms of topical raillery for the release of social tension from
interpersonal relationships. But they also gave expression to a
communal resistance to clerical domination that is at least partly
informed by an ancestral allegiance to alternative forms of
mediation with supernatural power. John Prim, a Kilkenny
antiquarian, hinted at the reality of this in 1853 in his cautious
description of the wake-games of his locality though he finally
judged the 'wake orgies' to be an anti-Christian and irreligious
phenomenon that deserved total suppression (Prim 1853: 334).
The 'Bull and the Cow', he writes,

> was another game strongly indicative of a Pagan origin, from
> circumstances too indelicate to be particularized. The game
> called 'Hold the Light' in which a man is blindfolded and
> flogged, has been looked upon as a profane travestie of the
> passion of our Lord; and religion might also be considered as
> brought into contempt by another of the series, in which a
> person caricaturing a priest, and wearing a rosary, composed
> of small potatoes strung together, enters into conflict with the
> 'Borekeen', and is put down and expelled from the room by
> direction of the latter.

Such mimes and imitative games are a feature of the merry wake that shows an uneven distribution pattern throughout the country. They are almost entirely absent in the mortuary tradition of the southwest corner of Ireland, the region most closely associated with cults and legends of the Hag/Goddess figures that underlie the *bean chaointe* and the mournful transition of the wake rituals. In the southwest, however, as everywhere throughout Ireland, the wake, whether unrestrainedly merry or not, is the occasion for another kind of creative and artistic performance. This centres on story-telling, with the wake-assembly being entertained throughout the night hours by the recital of hero tales, wonder tales, historical legends, memorates, genealogies, verse, proverbs, riddles, and other items of the major and minor genres of Irish oral narrative tradition. The wake-house is, in fact, one of the major loci for the (re)creative performance of this type of material, so that in this respect the wake-house resembles the 'rambling-house' or *teach áirneáin* (*áirneán* 'night-visiting'). Rambling-house is a term for those houses in each village and townland where during the winter half of the year from Hallowe'en to May Eve neighbours regularly gathered to listen to the histrionic performance of oral narrative material by local specialists who were regarded as excellent practitioners of this verbal art. In the wake-house, as in the rambling house, the verbal artistry was accompanied by a certain amount of singing, music-making, and dancing, and also by card-playing. During the dark hours of the nights over which the wake lasted, the assembly resorted in this fashion to that repertoire of verbal and other artistic and recreative performance that was regularly indulged in each year as a community activity of the rambling house, commencing with the coming of *Samhain*, Hallowe'en, and the onset of the winter half of the year. The period of six months from 1 November to 1 May was traditionally the season when the medieval bards taught their oral learning and craft in the bardic academies, and when the common people gathered, night after night, to listen to and witness and partake in the creativity of the rambling-house. *Samhain*, the great Celtic feast-day of the Otherworld and the dead ancestors, marks the beginning and initial high point of creative communal performance in the round of the traditional calendar. At the feast of *Samhain* the old Celtic year waned and the new year began, so that *Samhain*-time has a liminal, interlunary-like significance. It is a time when, in Irish tradition, the gateway between this life and the ancestral otherworld of the *sí* is wide open and household visitation from the *slua sí* ('fairy host') and from dead ancestors is expected to occur. Ordinary time and space are disordered at *Samhain*.

The Reeses (1961: 91) have emphasised this aspect of the *Samhain* festival:

A period of disorder in between the old year and the new is a common feature of New Year rituals in many lands but it is soon followed by the re-creation of an orderly world which lasts for another year. At Hallowe'en the elimination of boundaries, between the dead and the living, between the sexes, between one man's property and another's and, in divinations, between the present and the future, all symbolize the return of chaos. It is noteworthy that the 'day with a night' which the Mac Óc (a god-hero) equated with the whole of time were those of Samhain. This day partakes of the nature of eternity.

What Hallowe'en inaugurates is winter, and much of the uncanniness of night, when man seems powerless in the hands of fate, will prevail until the dawn of another summer. *Samhain* is the pivotal point in mythological time in Irish tradition. Great mythic events occurred at this time; while the first Gaelic Irish are said to have arrived on the island of Ireland at *Bealtaine* (May Day), the beginning of summer, the *Tuatha Dé Danann* ('the tribes or people of the Goddess'), who constitute the Celtic pantheon of Otherworld divines and semi-divines in Irish mythological tradition, had finally wrested the physical island of Ireland, and with it the hegemony of social existence from the Fomorions, the dark forces of chaos and disorder, at *Samhain*-time in the celebrated Battle of Moytura. This battle is the ultimate paradigmatic contest in Irish mythological tradition and cosmology.[8] It established the reign of the otherworld powers with whom the human Irish population was thereafter coexistent and who in later, Christianised times, became the *slua sí* ('the fairy host') of the imminent, contingent otherworld of ancestral tradition. There is a sense in which the same forces that opposed each other in the Battle of Moytura — the sovereign power of the representatives of the Goddess, on the one hand, and the disruptive and destructive force of death and disorder, on the other — are again brought into conjunction in the assembly of the merry wake. It is as if the paradigmatic, pivotal contest of mythological tradition is re-enacted in the passage ritual that marks the closest and gravest connections which the human realm has with the otherworld in ancestral cultural tradition in Ireland. In the traditional calendar, such closeness of worlds is symbolised and protected against in the festival of *Samhain* and the prophylactic performances to which it gives rise. The wake-assembly is a kind of movable *Samhain* similarly oriented to the conjunction of human and otherworld realms, and utilises similar performances of communal creativity to protect the human community from the dangers of contact with the immortal forces of the otherworld on whose threshold it is liminally poised within the ever-renewing cosmic contest between death and life.

The long survival of the Irish merry wake in the face of sustained clerical and middle-class opposition over the centuries is surely something explained in part by the way in which the merry wake functioned as an expression of a popular religious and cosmological sensibility that blended vivid native and localised apprehension of an otherworld and of death with the more intellectualised world-religion orthodoxy of Christianity. While there is not a single reference to any aspect of Irish tradition in van Gennep's work, *The Rites of Passage*, nevertheless the mortuary rituals of the Irish merry wake and funeral are consonant with his theories, not least in respect of his comment regarding the situation where, due to acculturation, more than one cosmological system informs the world view of a population (1960: 146): 'Funeral rites are further complicated when within a single people there are several contradictory or different conceptions of the afterworld which may become intermingled with one another, so that their confusion is reflected in the rites'. Despite such confusion, however, coherent significance derived from Celtic and pre-Christian cosmology endures in the central ritual of wake and funeral in later Irish popular tradition.

NOTES

1 Ó Súilleabháin (1967) quotes a variety of official ecclesi-astical pronouncements — synodal proclamations, pastoral letters, etc. — beginning with the Synod of Armagh in 1614 and ending with the Synod of Maynooth in 1927 in which Church authorities deplore and denounce various aspects of the merry wake.

2 While famine was endemic in Irish society both in medieval and early modern times, the years 1847–9 are referred to in Irish history as the years of the Great Famine.

3 Useful surveys of Celtic mythology and cosmology are provided by Mac Cana (1983) and Ross (1986). Important discussion of the nature and location of the native Irish otherworld is contained in Ó Cathasaigh (1978) and Carey (1983).

4 One such representative Goddess figure is that of the *Cailleach Bhéarra* (the Hag of Beare), who is known both in Gaelic learned medieval tradition and in the later oral narrative traditions of Ireland and Gaelic Scotland (see Ó Crualaoich 1988).

5 Ó hEochaidh (1977) is a corpus of oral narrative that includes many examples of such legends of fairy abduction from a single Donegal community, that of Teelin in the southwest of the county. The work also includes an introductory discussion by Máire Mac Neill of 'the connection between belief in the fairies and pre-Christian notions of the dead'.

6 In reporting this Ó Cróinín's informant expressed the belief that it was 'simply talk' rather than a prescription for how members of the community should behave in real life (Ó Cróinín 1980: 186). Nevertheless he said that he had himself been told of an occasion when this tactic had been tried in the face of what was believed to have been a case of fairy abduction. Other more tragic ordeals of burning and drowning are reported historically as having been occasionally inflicted on the living as a means of 'driving the fairy out of them' (see Jenkins 1977).

7 The beliefs and legends concerning the banshee in Irish folk tradition have been studied by Lysaght (1986).

8 Gray (1981, 1983) has most recently discussed the cosmological significance and mythological status of the texts of this tale which have survived. My own point regarding a possible symbolic connection between the myth of the Battle of Moytura and the ritual of the merry wake I judge to be supported by what Gray asserts regarding the continuing representative nature of ancient Irish mythic material for native Irish culture in later time:

> The transformation of Irish myth into pre-Christian history was the joint work of monastic scholars and the secular learned class, who together salvaged those traditions that were significant for the future as well as for the past. Even as a branch of native history, stories about the gods, the Túatha Dé Danann ('The Tribes of the Goddess Dana') remained paradigms for human action and models for the interpretation of human experience, although the gods themselves had become mortal descendants of Noah and contemporaries of Aeneas and Achilles. More specifically, such tales still served to establish legal precedent, to explain customary ritual, to justify political authority, to illustrate medical practice, and to define social order. [1981: 183]

REFERENCES

Carey, John (1983). The Location of the Otherworld in Irish Tradition. *Éigse* 19, 36–43.

Delargy, James H. (1946). The Gaelic Story-Teller. *Proceedings of the British Academy* 31, 3–47.

van Gennep, Arnold (1960). *The Rites of Passage,* trans. M.

V. Vizedom and G. L. Caffee. Chicago: University of Chicago Press.
Gray, Elizabeth A. (1981). *Cath Maige Tuired*: Myth and Structure (1–24). *Éigse* 18, 183–209.
—— (1983). *Cath Maige Turied*: Myth and Structure (24–120). *Éigse* 19, 1–35.
Hertz, Robert (1960). *Death and the Right Hand*, trans. R. and C. Needham. New York: Free Press.
Huntington, Richard and Peter Metcalf (1979). *Celebrations of Death*. Cambridge: Cambridge University Press.
Jenkins, Richard (1977). Witches and Fairies: Supernatural Aggression and Deviance among the Irish Peasantry. *Ulster Folklife* 23, 35–56.
de Latocnaye, Chevalier (1984 [1917]). *A Frenchman's Walk Through Ireland 1796–7*, trans. J. Stevenson. Belfast: Blackstaff Press.
Lysaght, Patricia (1986). *The Banshee*. Dublin: Glendale Press.
Mac Cana, Proinsias (1983). *Celtic Mythology*. London: Newnes Books.
Ó Cathasaigh, Tomás (1978). The Semantics of 'Síd'. *Éigse* 17, 137–54.
Ó Cróinín, Donncha (1980). *Seanachas Amhlaoibh Í Luínse*. Dublin: Irish Folklore Council.
Ó Crualaoich, Gearóid (1988). Continuity and Adaptation in Legends of *Cailleach Bhéarra*. *Béaloideas* 56, 153–78.
Ó hEochaidh, Seán *et al.* (1977). *Síscéalta Ó Thír Chanaill/ Fairy Legends From Donegal*. Dublin: Irish Folklore Council.
Ó Súilleabháin, Seán (1967). *Irish Wake Amusements*, trans. the author. Cork: Mercier Press.
Prim, John G. A. (1852). Olden Popular Pastimes in Kilkenny. *Journal of the Royal Society of Antiquaries of Ireland* 2, 319–35.
Rees, Alwyn and Brinley Rees (1961). *Celtic Heritage*. London: Thames and Hudson.
Ross, Ann (1986). *The Pagan Celts*. London: Batsford.

EMILY LYLE

Winning and Losing in Seasonal Contests

The focus of this study is on European tradition, although the parallel material drawn on to illuminate the problematic situation in Europe is also, of course, of interest in its own right. The question of what is involved in the winning and losing of seasonal contests may well be a key one, the answer to which will allow us to begin to sort out the bewildering tangle of seasonal traditions in Europe. The question has already been answered but the answer has been over-simplistic, relating only to pseudo-contests where the winning and losing sides are predetermined. For example, in the recent *Encyclopedia of Religion*, Theodor Gaster comments as follows under the heading 'Seasonal Ceremonies' (Eliade [ed.] 1987: 149):

The elimination of the old leads naturally to the inauguration of the new, that is, to rites of invigoration. The most widespread of these is the staging of a ritual combat between Fertility and Blight, Rainfall and Drought, Summer and Winter, or simply Life and Death, the positive protagonist (the one who personifies renewal) being always the winner.

Eliade had a special interest in pseudo-contests of this kind rather than real contests, and studied them exclusively (see, e.g. Eliade 1958: 319–21); Frazer, on the other hand, who gave many examples of seasonal contests in *The Golden Bough*, directed attention to the more complex situation that arises when either of the two sides may win.

Frazer referred to the battles between representatives of summer and winter which are found widely in Europe as mere dramatic performances since their outcome is unvarying (1911: 259). These battles occur at the beginning of summer and their action corresponds to the transition taking place in nature at that time, when winter can be said to 'yield' to summer, if we use a contest metaphor for the change of season. Frazer thought, however, that the dramatic shows or pageants familiar in Europe might have had their source in real contests where the outcome was uncertain, and pointed to Eskimo parallels where the representatives of summer and winter engage in a tug-of-war in which either side may win. Frazer had no European cases of real contest between the seasons to draw on, but it is now possible to adduce an instance which apparently stems from Europe — the narrative of the contest between Oenomaus and Pelops as told in the sixth-century Byzantine chronicle of John Malalas.[1] This can be taken as a starting point in the present enquiry.

According to this account, Oenomaus, king of Pisa, holds a chariot-race on 25 March each year in which he himself is one of the contestants and his opponent is a stranger from another country. The loser of the contest is put to death. The two opponents contend both as individuals and as representatives of halves of the community. The stake for the individual is his life but I am not concerned with exploring the level of the individual here. The enquiry is directed towards what the victory of one and the defeat of the other means for the halves of the community each represents.

The two opponents cast lots to determine which of them shall represent which half of the community. The two are a 'blue' half (whose representative wears blue), which relates to Poseidon and the sea; and a 'green' half (whose representative wears green), which relates to Demeter and the land. Sailors and people living on the coast desire the victory of the blue contestant, while peasants and those living inland desire the victory of the green contestant. It is held that if the green contestant wins there will be a dearth of fish, and that if the blue champion wins there will be a dearth of grain and fruit, wine, and olive oil. This is to put it negatively, in terms of the fears of the two parts of the community. Put in positive terms, a victory by the blue half indicates that there will be abundant fish, and a victory by the green half indicates that there will be abundant grain, fruit, wine, and olive oil. The success or failure of each half in securing an abundant supply of the produce appropriate to it over the period of a year relates to the success or failure of each half in the annual contest. It is a zero-sum situation, where the win by one half is a loss by the other.[2]

There is no mention of seasons in the account by Malalas, but the colours blue and green are understood by him as those of the circus factions, and the faction colours are related to the seasons in Malalas and elsewhere. The precise correspondence between the four faction colours — blue, white, green, and red — and the four seasons of spring, summer, autumn, and winter is still a matter for debate, but, by the connections I have formerly proposed, the half including blue corresponds to the winter half of the year, and the half including green to the summer half.[3] Although the temporal aspect is merely latent so far as the narrative is concerned, the two halves can be seen to be both spatial and temporal, with one being *land or land lying inland* and *summer*, and the other *sea or coast* and *winter*.

So far as is known, the narrative in Malalas does not have reference to any contemporary ritual — except that the blue and green factions were contending vigorously in the hippodrome — and we lack context for the contest. The fears of the onlookers, however, appear to be appropriate to a society with division of labour and a market economy, where a scarcity of fish would particularly affect fishermen and a crop failure would particularly affect peasants. For a contest in a society without division of labour apart from that between the sexes, we can turn to the tug-of-war among the Inuit (formerly called Eskimo) of Cumberland Sound in Northern Canada referred to by Frazer, where the contest of the halves involves all the men of the community.

The account by Franz Boas drawn on by Frazer in his main text is as follows (Boas 1888: 605):

> The crowd [of men] next divides itself into two parties, the ptarmigans (axigirn), those who were born in the winter, and the ducks (aggirn), or the children of summer. A large rope of sealskin is stretched out. One party takes one end of it and tries with all its might to drag the opposite party over to its side. The others hold fast to the rope and try as hard to make ground for themselves. If the ptarmigans give way the summer has won the game and fine weather may be expected to prevail through the winter (nussueraqtung).

Here a victory by summer means that the winter will be a good one, but Frazer refers to another account which Boas received from James S. Mutch, where it is the victory of winter that means that there will be plenty of food that winter (Boas 1907: 140–1):

> Then the people take a long rope, the ends of which are tied together. They arrange themselves so that those born during the summer stand close to the water, and those born in the winter stand inland; and then they pull at the rope to see whether summer or winter is the stronger. If winter should win, there will be plenty of food; if summer should win, there will be a bad winter.

The two different interpretations given both prognosticate a good
winter, and, if they are accurately recorded, may either reflect
different methods of interpretation in the two communities or,
conceivably, may be alternative interpretations available within a
single community so that there would always be hope of a good
winter through one interpretation or the other.

In the second account a spatial relationship is mentioned. The
summer people ('ducks' in the first account) stand on the sea side
and the winter people ('ptarmigans' in the first account) stand
inland. Whatever interpretation should be put on this, it is clear
that, in the broader framework of the Inuit seasonal pattern, the
produce of winter is the sea mammals of the coast and the produce
of summer is the land animals, notably caribou, which are hunted
inland, so that a good summer would mean abundance of the meat
of land animals and a good winter would mean abundance of the
meat of sea animals (Mauss 1979; Pearce 1988). In this society, all
the men are engaged in hunting sea animals and land animals
according to season. If the people of winter win the tug-of-war, the
whole community has greater abundance in the winter; and,
presumably, if the people of summer win, the whole community
has greater abundance in the summer.

In the Inuit instances, we have correspondences between time,
place, and product, of which the temporal register is made explicit
in the accounts of the contests, which are between summer people
and winter people; and the other connections are necessary ones
when one knows where the people live in the two seasons and what
their food sources at these times are. The correspondences are
summer, land, and *meat of land animals*; as against *winter, sea*,
and *meat of sea animals*. This reflects the heavy reliance on meat
found among the Inuit living in an extreme climate. In the Malalas
passage, taken together with the seasons indicated by the circus
colours, the correspondences are: *summer, land*, and *crops*; as
against *winter, sea*, and *fish*.

The tie between the result of an annual contest and the relative
abundance of different foods in the coming year occurs in a North
American instance discussed in a recent study of sport by David
Sansone. This contest is the relay race held among the Jicarilla
Apache where 'one "side" represents the interests of the food
animals and the other side the interests of vegetation, and the
victory of one side or the other ensures that, for the ensuing year,
the supplies of meat will be greater than those of fruits and
vegetables or vice versa' (Sansone 1988: 14). In this case, there is
a clear division of the two types of product: meat as against fruit

and vegetables. The distinction is fully recognised by those involved in the race; for example, the kilts of the principal runners on one side are painted with animals, including the deer, while the kilts of the principal runners on the other side are painted with plants and fruits such as maize and chokeberries (Opler 1944: 82, 88–9). Seasons and places as well as products are represented in the relay race of the neighbouring Pueblo peoples, from whom the Jicarilla Apache race was a cultural borrowing,[4] and here, for the first time in this discussion, we have the opportunity to look at a complete *season, place*, and *product* series in a rich ethnographic context which puts the existence of the correspondences — in this particular instance — beyond doubt.

Among the Pueblo Indians, sun/earth is the basic cosmological pair, and 'the next basic level of duality is that in nature, winter and summer, providing the fundamental principle of organization for the ritual calendar' (Ortiz [ed.] 1972: 144). In what follows, I shall refer especially to the Tewa group of Pueblo Indians; they have been the subject of an outstanding study by Alfonso Ortiz, who has paid detailed attention to their cosmology. Among the Tewa, east and north relate to winter, and west and south to summer. The actual sides in the relay race are referred to as those of north and south, but the tradition is that the sides used to be those of winter and summer (Ortiz 1969: 108–11), and, on the day of the races, 'the North moiety carries a red and white standard which the Winter moiety chief has decorated with fox skins, while the South moiety carries a yellow and blue standard which the Summer moiety chief has decorated with cottonwood branches' (Ortiz 1969: 109). The three points of season, place, and product are all present together in a passage in a Tewa origin myth (Parsons 1926: 15):

> The eastern mountains belonged to the Winter People and the western mountains belonged to the Summer People. Those on the Eastern Mountains, the Winter People, were eating deer, elk (the meat of wild animals); and the Summer People were eating . . . all different kinds of fruit.

Each of the two chiefs rules the village for a part of the year, 'the Winter Chief from about the autumnal equinox until a month before the vernal, and the Summer Chief for the remainder of the year' (Ortiz 1965: 390). The divisions of the year relate to subsistence activities, the transfer to the Summer Chief being seen as initiating the agricultural season, and the transfer to the Winter Chief ushering in a 'period of intensified hunting, which begins after all of the late crops are in storage' (Ortiz 1965: 319). Each

chief initiates an infant into the appropriate moiety by giving the
child either a portion of dried meat, in the case of the Winter
moiety, or a sweet drink made from sprouted grain, in the case of
the Summer moiety (Ortiz 1965: 391). As can be seen, among the
Tewa the connections between seasons and products are clear and
explicit.

A return can now be made to Europe, where it is possible also to
find annual contests the results of which are held to affect the
produce of the coming year. Since these contests are not specified
as being between summer and winter, they occur at a different
point in *The Golden Bough* from the ones mentioned above, and
appear, rather inappropriately, under the notion of the expulsion
of evil. In the contests in which good is ranged against evil, good is
meant to win, as Frazer remarks (1913: 180–1), but some of the
instances he gives do not fit among these biased contests,
belonging instead to the present discussion. In Normandy, for
example, several hundred players struggled for a ball on Shrove
Tuesday, with the aim of lodging the ball in the home parish and 'it
was thought that the parish which was victorious in the struggle
would have a better crop of apples that year than its neighbours'
(Frazer 1913: 183). Frazer suggested that popular football in
Britain should be seen as a contest of the same kind (1913: 184),
and John Robertson has since shown a connection between the
outcome of a football match as played in Orkney and the
anticipation of abundance in the coming year; in this case, there
are two different products, a point that was lacking in the case
from Normandy.

The two sides are called Uppies or Up-the-Gates and Doonies
or Doon-the-Gates, and they attempt to carry the ball respectively
either to the Up goal inland or to the Down goal, which is the
harbour. Robertson records that 'an old belief in Kirkwall was that
a win for the Up-the-Gates brought heavy crops and a good
harvest, whereas successful fishing followed a Down-the-Gate
victory' (1967: 93). For twenty-nine years from 1846 the ball in the
New Year's Day game had gone Down, but the Uppies won on
New Year's Day in 1875, and Robertson notes: 'Later that day an
old man was heard to remark that as it was 1846 when the potato
blight first appeared in Orkney, "we'll surely hae guid tatties this
year, after the ba' has gaen up."' Robertson adds (1967: 93–4):

An old rhyme goes:
> Up wi' the ba' boys,
> Up wi' the ba',
> An' ye'll get cheap meal,
> An' tatties an' a'.

On the other hand for the Down-the-Gates, success brought the promise of an abundance of herring. So we have here two distinct locations, land (up) and sea (down), and two distinct products, crops (grain and potatoes) and fish, as in Malalas. The scheme does not relate clearly to the seasons,[5] but it can be suggested that in an earlier scheme (of which we have reminiscences in the battle between summer and winter) one of two halves of the year was associated with one of two locations and with one of two valued types of product. If the cosmological scheme envisaged did once exist, it must be understood to have split apart. When the two halves of the year were thought of in isolation from places or products, the customary contest would tend to turn into a pageant in which summer defeated winter, since the evident relationship between the two halves of the year was that one displaced the other. A sense of paired places could have continued independently of a sense of paired times, and Dodgshon has outlined how it could have remained in a non-cosmological era following the 'disaggregation of the scheme' (Dodgshon 1985: 64);[6] certainly, customary annual contests between places exist that are unrelated to seasons and products. In these, in some undefined way, it is held to be lucky to win the contest between the paired places. This is potentially a more divisive idea than it would have been in a cosmological society where the win by either side was equally beneficial to the whole community; only the kind of benefit varied. The converse lack, of course, would also have been shared throughout the community.

The Orcadian instance and the one in Malalas are so close that they can be put together and may tentatively be identified as a 'European' scheme, shown in the Table as A. The stress on fish here should perhaps be taken as an indication of the importance of the dry/wet contrast, which might conceivably in other contexts take other forms, such as a contrast between food and drink. Clearly, meat had a place in the economy, but it remains open whether it should be understood to combine with fish or with crops. The Pueblo lay stress on meat, but, as fish scales are ritually scattered by the Winter Chief while the Summer Chief scatters cornmeal (Ortiz 1969: 115), it is possible to say that fish is included with meat in the winter half. The Pueblo scheme is shown as B, and includes the foods as paired by the Jicarilla Apache, who, as Opler notes, being primarily hunters and gatherers, had 'a profound interest in a dual food supply' consisting of meat and of uncultivated growing things (Opler 1944: 97). The distinctive Inuit scheme is given under C.

Table

A	
Summer	Winter
Land	Sea
Fruit and crops	Fish

B	
Summer	Winter
South and west	North and east
Fruit and crops/vegetables	Meat (fish)

C	
Summer	Winter
Land	Sea
Meat of land animals	Meat of sea animals

The results of the contest are related to the results in obtaining sustenance in the course of the year, but it is possible to read the relationship in at least three different ways, which can be listed as follows (with examples taken from a win by the winter side in scheme A). The relationship could be (*i*) causal — because winter wins there will be more fish; that is, it is possible to affect the course of the year by making a winning effort in the contest; or (*ii*) divinatory — since the year holds the event that there will be more fish, the contest will show this in advance; or (*iii*) metonymic — the contest and the year have the same nature and this nature becomes evident in the result of the contest; the part is a sample of the whole. In cases (*ii*) and (*iii*), the contest could be replaced by a drawing of lots.

As we enter the relatively little-known field of the cosmology of Europe with the aim of straightening out the tangles created by the criss-crossing of the multiple threads of tradition over long periods in a post-cosmological era, the study of the annual contest that relates to the products of the seasons may allow us to make a useful start. In the European context, this contest lends itself to being negatively defined — it is not a biased contest in which good inevitably defeats evil, and it is not a sham contest providing a dramatic presentation in which the new season defeats the old. It is a real contest in which both sides have a chance of winning, and in which it appears that the success of either side is equally desirable.[7]

Before concluding, it may be worth looking at what the comparative evidence suggests about how we should regard the time of year of the contest. Ideally, it would seem that the contest should take place at a year-beginning and that this should coincide with the start of one of the halves of the year, but it is clear from the comparative evidence that it is the concept of year-halves that is primary and that the actual date of the contest is of lesser importance.

The Inuit of Cumberland Sound had such a strongly marked divide at the equinoxes in March and September that they held that the world turned over at these times, so that what was above in summer was below in winter and vice-versa (Boas 1907: 130–1). However, the tug-of-war witnessed by Boas in 1883 took place on 10 November, and he notes that three years later it took place on the same date (1888: 669). The necessary ingredients appear to be simply a division into summer and winter, and a contest that relates to their differing fortunes. If the contest was once tied to a point of division between the seasons, it has since strayed to another point of the year. This point, of course, can still have a significance as the time when the ice-floes form (cf. Pearce 1988: 316) without marking one of the pair of transitions between year-halves which can apparently be identified with the equinoxes.

Among the Pueblo, the contests do take place at a transition between the year-halves, but the periods covered by the two halves vary. Ortiz notes that 'at Hopi and Zuni the transition is determined by the solstices, while for the rest it seems to be by the equinoxes, either actual or as culturally construed' but that, regardless of these differences, 'the duality is still winter and summer' (1972: 144).

In the case of Europe, we find both wide variations in dates of contests and also a variety of possible year-beginnings. This study indicates that we can usefully approach the problems of the ritual calendar through trying to grasp the meaning of the seasons in terms of a correspondence system; and that we will do well to hold firmly in mind in the midst of all the flux and variation the probability that the duality expressed through this annual contest is 'still winter and summer'.

NOTES

1 Trans. Jeffreys *et al.* (1986: 92–3); the passage is discussed in Lyle (1990).
2 This situation is defined by Lester C. Thurow in the following passage (1981: 11):
    A zero-sum game is any game where the losses exactly

equal the winnings. All sporting events are zero-sum games. For every winner there is a loser, and winners can only exist if losers exist. What the winning gambler wins, the losing gambler loses.

3 In this scheme, autumn corresponds to green and the female, and winter to blue/black and the cultivators; see further Lyle (1990).

4 Opler (1944: 91–6). For map showing the location of the Jicarilla Apache and the Pueblo (see Dozier 1970: frontispiece).

5 A possible link with the two seasons may be suggested by an Orkney legend of the good Sea Mither and the evil Teran fighting at each equinox for control of the sea, with the Sea Mither gaining dominance for the summer period and Teran for the winter months, but, although Robertson quotes it, he does not claim that a connection with the game can be made out. Cf. also the story of Njord and Skadi in Scandinavian mythology (Davidson 1964: 30, 106–7).

6 Cf. the retention among the Xerente of Brazil of the memory of the dualistic properties of their traditional culture even nowadays when 'people are allocated to the two sides on an ad hoc basis' at the time of ceremonies such as log races (Maybury-Lewis 1989: 108–9, cf. 102–3).

7 In the exceptional conditions among the Inuit, however, as seen above, the win that relates to a good winter is apparently preferred. Cf. Fox (1979: 165–71, especially 170) (Savu, Indonesia), and Yoshida and Duff-Cooper (1989: 213) (Bise, Okinawa, Japan) for instances where, although either side may win in a contest, the win by one particular side is the more auspicious.

REFERENCES

Boas, Franz (1888). *The Central Eskimo*. Washington: Sixth Annual Report of the Bureau of Ethnology.
—— (1907). *The Eskimo of Baffin Land and Hudson Bay*. New York: Bulletin of the American Museum of Natural History 15.
Davidson, H. R. Ellis (1964). *Gods and Myths of Northern Europe*. Harmondsworth: Penguin.
Dodgshon, Robert A. (1985). Symbolic Classification and the Development of Early Celtic Landscape. *Cosmos* 1, 61–83.
Dozier, Edward P. (1970). *The Pueblo Indians of North America*. New York: Holt, Rinehart and Winston.
Eliade, Mircea (1958). *Patterns in Comparative Religion*. London: Speed and Ward.
—— (ed.) (1987). *The Encylopedia of Religion*. New York and London: Macmillan and Collier Macmillan.

Fox, James J. (1979). The Ceremonial System of Savu. In *The Imagination of Reality: Essays in Southeast Asian Coherence Systems*, eds A. L. Becker and Aram A. Yengoyan, pp. 145–73. Norwood, New Jersey: Ablex.

Frazer, James George (1911). *The Golden Bough*, Part 4 *The Dying God*. London: Macmillan.

—— (1913). *The Golden Bough*, Part 6 *The Scapegoat*. London: Macmillan.

Jeffreys, Elizabeth *et al.* (trans.) (1986). *The Chronicle of John Malalas*. Melbourne: Australian Association for Byzantine Studies, Byzantina Australiensia 4.

Lyle, Emily (1990). *Archaic Cosmos: Polarity, Space and Time*. Edinburgh: Polygon.

Mauss, Marcel (1979). *Seasonal Variation among the Eskimo*, trans. from the French by James F. Fox. London: Routledge and Kegan Paul.

Maybury-Lewis, David (1989). Social Theory and Social Practice: Binary Systems in Central Brazil. In *The Attraction of Opposites: Thought and Society in the Dialectic Mode*, eds David Maybury-Lewis and Uri Almagor, pp. 97–116. Ann Arbor: University of Michigan Press.

Opler, Morris Edward (1944). The Jicarilla Apache Ceremonial Relay Race. *American Anthropologist* 46, 75–97.

Ortiz, Alfonso (1965). Dual Organization as an Operational Concept in the Pueblo South-West. *Ethnology* 4, 389–96.

—— (1969). *The Tewa World*. Chicago: University of Chicago Press.

—— (ed.) (1972). *New Perspectives on the Pueblos*. Albuquerque: University of New Mexico Press.

Parsons, Elsie Clews (1926). *Tewa Tales*. New York: G. E. Stechert. Memoirs of the American Folk-Lore Society 19.

Pearce, Susan M. (1988). Ivory, Antler, Feather and Wood: Material Culture and the Cosmology of the Cumberland Sound Inuit, Baffin Island, Canada. *Cosmos* 4, 307–21.

Robertson, John (1967). *Uppies and Doonies: The Story of the Kirkwall Ba' Game*. Aberdeen: Aberdeen University Press.

Sansone, David (1988). *Greek Athletics and the Genesis of Sport*. Berkeley, Los Angeles, and London: University of California Press.

Thurow, Lester C. (1981). *The Zero-Sum Society: Distribution and the Possibilities for Economic Change*. Harmondsworth: Penguin.

Yoshida, Teigo and Andrew Duff-Cooper (1989). A Comparison of Aspects of Two Polytheistic Forms of Life: Okinawa and Balinese Lombok. *Cosmos* 5, 213–42.

Notes on Contributors

ELIZABETH BAQUEDANO is a part-time lecturer at Birkbeck College, University of London. Since completing her doctoral thesis on Death in Aztec Sculpture at the Institute of Archaeology, University of London, in 1989, she has been running the Fifth Centenary Office 1492–1992 for the United Kingdom. Her main research interests are in the field of Aztec Sculpture.

GLENYS DAVIES is a lecturer in Classical Archaeology in the Department of Classics, University of Edinburgh. Her main research interests are in the field of Roman funerary imagery. She is currently Treasurer of the Traditional Cosmology Society, and edited *Cosmos* 5 (1989) ('Polytheistic Systems').

ANDREW DUFF-COOPER was the 1989–90 Cosmos Fellow of the University of Edinburgh and is now a Professor in the Department of Humanities, Seitoku University, Japan. He has published widely on Balinese life in Pagutan, western Lombok. Until recently a visiting lecturer at Keio University, Tokyo, he has also published articles about aspects of Japanese ideology. This work continues.

JOY HENDRY is a reader at the Scottish Centre for Japanese Studies, University of Stirling, and a principal lecturer at Oxford Polytechnic. She has carried out fieldwork in Japan on several occasions since 1975, and is the author of *Marriage in Changing Japan* (1981), *Becoming Japanese* (1986), and *Understanding Japanese Society* (1987). She is currently working on politeness behaviour in Japan.

JANET HOSKINS is Assistant Professor of Anthropology at the University of Southern California, Los Angeles. She has done extensive fieldwork in Kodi, West Sumba, and has published articles on traditional poetics, art, gender, head-hunting, conversion, and emerging historical consciousness. She is presently writing a monograph on feasting, exchange, and temporality in Kodi, and is collaborating on several ethnographic films.

ALICE B. KEHOE was educated at Barnard College (B.A., 1956) and Harvard University (Ph.D. 1964), majoring in Anthropology at both institutions. She taught Anthropology at the University of Regina (1964–65) and the University of Nebraska (Lincoln) (1965–68), and since 1968 at Marquette University in Milwaukee, Wisconsin, where she is now Professor of Anthropology. Her principal fieldwork has been in both Archaeology and Ethnography among the Blackfoot, Plains Cree, and Saskatchewan Dakota of the Northwestern Plains (Montana, Alberta, Saskatchewan). Her textbook, *North American Indians: a Comprehensive Account* (Prentice-Hall, 1981) is widely used, and she recently published *The Ghost Dance: Ethnohistory and Revitalization* in the Holt, Rinehart and Winston series 'Case Studies in Cultural Anthropology'.

KATHRYN LOWRY is a doctoral student in the Department of East Asian Languages and Civilizations at Harvard University, researching Chinese oral poetry and song. She has published results of interviews with artists and musicians trained in the People's Republic of China, as part of a comparative study of arts education done at Harvard Project Zero.

EMILY LYLE is a research fellow at the School of Scottish Studies, University of Edinburgh, and is currently President of the Traditional Cosmology Society. Her publications include *Archaic Cosmos: Polarity, Space and Time* (1990).

GEARÓID Ó CRUALAOICH is a lecturer in Folklore at University College, Cork, of the National University of Ireland. He has been a postgraduate student of Anthropology at the University of Pennsylvania and the London School of Economics. During 1984 he was a fellow at the Institute for Advanced Studies in the Humanities at the University of Edinburgh. He has published various articles and book chapters in both Irish and English on literary, linguistic, and ideological aspects of Irish tradition. Currently he is researching the question of how female cultural stereotypes impinge on the lives of women in rural Irish society.

D. L. SHARMA, born into a family of Brahmin priests in India in 1911, was educated at the Allahabad and Agra Universities, and at the University of Edinburgh. He has lectured at many colleges and universities in Britain, and since 1979 has published six books and various articles on topics connected with Hinduism. This work continues.